Radiobiology and
Bio-Medical Research

Radiobiology and Bio-Medical Research

Editor

K.P. Mishra

Narosa Publishing House
New Delhi   Chennai   Mumbai   Kolkata

EDITOR
K.P. Mishra
Head
Radiation Biology and Health Sciences Division
Bhabha Atomic Research Centre
Mumbai, India

This book is published for The Society for Cancer Research and
Communication (SCARC), P.O. Box 8094, Mumbai 400 056, India

ISBN 81-7319-484-X

Published by N.K. Mehra for Narosa Publishing House, 22 Daryaganj,
Delhi Medical Association Road, New Delhi 110 002 and typeset by
Replika Press, Pvt. Ltd. and printed at Baba Barkha Nath Printers,
New Delhi, India.

This volume is dedicated to my revered parents
Shri Ram Narain Mishra
and
Late Smt Shivakali Devi Mishra

Foreword

Radiation has existed since the origin of this planet. Radioisotopes and ionizing radiation have found increasing applications in industry, agriculture and medicine. Understanding the effects of ionizing radiation on cells and tissues has attracted scientists from multidisciplinary research backgrounds. Man made radiation sources act as double edged weapons both for useful and harmful purposes. Extensive research efforts in the past decades have provided several useful technologies. With the development of science and technology, various novel and unique applications are emerging. Diagnosis and treatment of human cancer by radiation are some notable examples, which have proved beneficial to society.

With the development of advanced biophysical techniques and newer molecular approaches, understanding of the basic mechanisms of the effects of ionizing radiation on cells and tissues including humans, seem to be in sight. This volume has a wide coverage of health and radiation related scientific material identifying prospects and challenges. Written by experts, scientific content of each topic presents a state-of-the-art scenario addressing critical issues.

Though it is difficult to cover all aspects of this ever-developing field, the present volume certainly serves as a guideline for the current trends and future directions in radiology and biomedical research.

I am confident that this book would prove invaluable and useful to a wide spectrum of professionals including specialists, educationalists and policy makers.

GIRJESH GOVIL
SENIOR PROFESSOR

Preface

The radiation biology and medicine are closely related branches of science attracting scientists from multi-disciplines such as biophysics, radiation chemistry, medicine, biotechnology and health sciences. Convergence of biology and medicine was witnessed over a period of time in the era of modern science but the prospects of marriage between radiation biology and medicine became evident soon after the discovery of X-ray and radioactivity at the end of 19th century. Research progress during the past decades has made substantial contribution to the understanding of radiobiological principles on the basis of which greater benefits have been derived to human health through practice of medicine. This has essentially been the basis of inspiration to compile this volume, which spans current developments in radiobiology and emerging biomedical applications related to human health, such as, antioxidants, nuclear diagnostics and cancer radiotherapy.

The biology of ionizing radiation effects has continued to remain fascinating area of research. Over the years, rapid progress has been made in applying radioisotopes to medical problems and in understanding the effects of ionizing radiation on vital cellular molecules, cells and tissues including effects on animals and humans. For example, cytotoxicity effect of radiation has been exploited in cancer radiotherapy but the dream of killing cancer cell and sparing the normal cell remains unrealized. A number of radiation technologies, namely, nuclear reactors for electricity, radiation sterilization of foods, nuclear diagnostics and cancer radiotherapy, radition methods for developing new variety of plants, animals and crops are some of the examples rendering service to society. However, further basic research is warranted to develop newer and more effective technologies for improving the quality of human life. This volume is a collection of recent results on some of the frontline topics of biomedical and radiobiological research. Each Chapter has been meticulously written by practicing scientists with established expertise in the specific field and, therefore, the book is intended to serve a source of latest information in the particlar field and, therefore, the book is intended to serve a source of latest information in the particular field of research and technology. It is hoped that this volume would prove useful and handy to both, young researchers and established investigators in biomedical science.

I am indeed grateful to contributors of the articles. I wish to acknowledge the President and Office Bearers of Indian Science Congress Association providing me an opportunity to organize symposium in my role as the Sectional President, which prompted me to compile this volume from the presentations in the 88th Session of Indian Science. I have a special appreciation of the encouragement and support of Prof B.P. Chatterjee for academic inspirations. The support of Dr N. Huilgol through Society of Cancer Research and Communication (CRAC) is the ultimate drive and motivation in making this volume possible. Among others, the support of Prof T. Nomura of Osaka University, Japan deserves a special commendation. I owe fully to my colleagues, especially Shri B.N. Pandey and my doctoral students, who readily and willingly helped me in editing and organizing the scientific material. Lastly, I express my appreciation and gratitude to my wife, Usha and my children, Rajesh and Anil, for their patience and silent support in this endeavor without complaining of my absence from home for late hours of working including on holidays.

K.P. MISHRA

Contents

Oxidative Stress, Radiation and Health

Radiobiology and Bio-Medical Research
Edited by K.P. Mishra
Copyright © 2004 Narosa Publishing House, New Delhi, India

1. Antioxidants in Health and Disease

Dipak K. Das and Nilanjana Maulik

Cardiovascular Research Center, University of Connecticut School of Medicine,
Farmington, CT 06030, USA

Abstract The role of antioxidants in health and disease is now well recognized. Antioxidants reduce or eliminate the cytotoxic reactive oxygen species (ROS) that are generated in the tissues in pathophysiological conditions. Biological tissues maintain a critical balance between antioxidant reserve and ROS, which are continuously produced in the cells. In pathophysiological conditions, this balance is upset because of reduction in antioxidant reserve and excessive production of ROS, which play a crucial role in many degenerative diseases including cardiovascular diseases such as ischemic heart disease, atherosclerosis and congestive heart failure. The potential for antioxidant therapy, therefore, continues to grow at a staggering pace.

The results of our study indicate that myocardial defense against heart diseases comprises several lines of defense. The first line of defense consists of intracellular antioxidants such as superoxide dismutase, catalase, glutathione peroxidase and glutathione reductase while the second line of defense contains lipolytic and proteolytic enzymes as well as proteases and phospholipases. Recently, we have identified a third line of defense consisting of oxidative stress-inducible genes and proteins which are produced by the heart in an attempt to counteract the invading ROS in an inducible pathway regulated by a precise signal transduction system. We have identified an intracellular signaling cascade regulated by tyrosine kinase phospholipase coupled activation of MAP kinases and protein kinase C. The redox signaling controls ROS function as second messenger in potentiating this signaling cascade and generation of antioxidants in this inducible pathway. Several redox-sensitive transcription factors including NF_kB and AP-1 as well as redox regulated genes such as Bcl-2 play a key role in the adaptive synthesis of intracellular antioxidants triggered by redox signaling of ROS.

The present report focuses on the signaling mechanisms of the expression of ROS-induced genes for these intracellular antioxidants and their transcription regulation. The results of our studies documented that these antioxidants generated as an adaptive response against stress-induced cardiovascular diseases play an important role in maintaining the balance between antioxidant reserve and ROS and, thus, dictates the need of antioxidant supplementation to combat against these degenerative diseases.

Introduction

A large number of degenerative diseases, including coronary heart disease, have been linked with the overproduction of oxygen-derived free radicals. The results of many studies, including our own, have demonstrated that excessive production of reactive oxygen species in concert with drastic reduction of antioxidant reserve play a crucial role in the pathophysiology of ischemic heart disease [9]. Myocardial ischemia and reperfusion cause the cardiomyocytes to face conditions that shift their redox status to undergo a drastic change subjecting them to oxidative stress. Interventions with oxygen free radical scavengers or antioxidant therapy have been found to be cardioprotective against ischemic reperfusion injury [11].

Recent studies have documented that coronary heart diseases cause cardiomyocyte death not only by necrosis but also by apoptosis [3]. Reperfusion of ischemic myocardium results in apoptotic cell death and DNA fragmentation. In concert, ischemia/reperfusion is associated with the induction of a number of both pro and anti-apoptotic genes and transcription factors [5].

Three Lines of Antioxidant Defense

Reactive oxygen species are continuously generated in the heart during incomplete reduction of oxygen molecule. These include superoxide anions and hydroxyl radicals as well as oxidants like hydrogen peroxide. These reactive oxygen species lead to the production of a number of secondary free radicals including peroxyl and hydroperoxyl radicals. Nitric oxide generated in the tissue can produce extremely cytotoxic peroxynitrites. Virtually every biomolecules including lipids, proteins and DNA molecules are potential targets of free radical attack. Mammalian heart is equipped with its own antioxidant defense to fight any free radical attack [2]. Constitutive cellular protection against acute stress can be provided by various intracellular antioxidants such as glutathione, α-tocopherol, ascorbic acid, β-carotene and antioxidant enzymes that include superoxide dismutase, catalase and glutathione peroxidase. These cellular compounds reduce/eliminate the oxidative stress by directly quenching the reactive oxygen species before they damage vital cellular components and therefore, can be considered as part of the *first line of defense* against the external stress. Often, because of the inadequacy of the intracellular antioxidants or due to the presence of increased amount of oxidative stress in pathophysiologic conditions such as ischemia and heart failure, the reactive oxygen species may reach their targets that include nucleic acid, protein and lipids. This results in the injury to the cellular components causing DNA strands to break, protein degradation and lipid peroxidation. Mammalian hearts are also protected by a *second line of defense* system consisting of several lipolytic and proteolytic enzymes, proteases and phospholipases that are involved in the systematic recognition and removal of injured components.

In view of the increasing evidence for the involvement of cellular oxidation and oxidative stress in various pathophysiologic conditions of heart disease, it is of considerable interest to examine whether myocardial cells possess an inducible pathway for the antioxidant defense. The signal transduction pathways by which various stress signals are translated into oxidative stress leading to the modulation of antioxidants/antioxidant enzymes are likely to be different, but the induction of the expression of their mRNAs and/or the increased translation of mRNA accumulation as a result of changes at the level of RNA transcription or stability seem to be the most plausible mechanism. This could lead to the synthesis of proteins involved in cellular protection or repair of injury. These phenomena related to specific gene expression could be viewed as a *third line of defense,* and may reflect the ultimate adaptive responses.

Antioxidant Protection of the Heart

It should be clear from the above discussion that despite the presence of adequate antioxidant defense, hearts becomes helpless in pathologic conditions because of the reduction of antioxidants in conjunction with the massive production of reactive oxygen species. There is a general agreement that the amount of several antioxidants and antioxidant enzymes are significantly reduced after ischemia and reperfusion. For example, reduced amount of SOD, catalase and glutathione peroxidase enzymes as well as α-tocopherol and ascorbic acid have been found in the ischemic reperfused myocardium. The loss of the key anti-oxidant enzymes and antioxidants thus reduces the overall antioxidant reserve of the heart and make the heart susceptible to ischemia/reperfusion injury. Thus, reduced antioxidant reserve in conjunction with increased free radical generation leads to the cellular injury associated with ischemia/reperfusion.

Antioxidant supplementation to the heart in the face of ischemia has been found to be cardioprotective. Treatment of hyperthyroid rat with α-tocopherol protected the hearts from ischemic reperfusion injury. Antioxidant therapy with SOD plus catalase prevented apoptotic cardiomyocyte death induced by ischemia and reperfusion. Selenium deficiency led to an increase in myocardial ischemic injury to the rats. IRFI 042, a novel vitamin E-like antioxidant inhibited activation of NFkB and reduced inflammatory response in myocardial ischemia/reperfusion injury. Grape seed proanthocyanidins protected rat hearts from ischemic injury by scavenging hydroxyl and peroxyl radicals generated during ischemia and reperfusion. In another related study, the red wine antioxidant resveratrol protected isolated rat hearts from ischemia reperfusion injury [2]. Ischemia/reperfusion-induced oxidative stress was found to be reduced by vitamin E treatment. Recombinant adenovirus-mediated cardiac gene transfer of SOD and catalase attenuated postischemic contractile dysfunction. Another study showed that overexpression of MnSOD protected against myocardial ischemia/reperfusion injury in transgenic mice. Pretreatment of the heart with SOD plus catalase was also found to ameliorate ischemic reperfusion injury. Thus, the potential for antioxidant therapy to combat ischemic heart disease appears to be enormous.

Micronutrients, Antioxidants and Heart

Micronutrients are essential for survival because they play a key role in metabolism. A growing body of evidence indicates a crucial role of several micronutrients in cardioprotection. Of particular importance are vitamins A, C and E, folate, β-carotene, coenzyme Q_{10}, zinc and selenium because absence of one or more of these micronutrients has been found to comprise important risk factor for cardiac diseases [10].

Clinical use of antioxidant vitamin supplementation may help to prevent coronary heart disease. Beneficial effects of vitamin E treatment in acute myocardial infarction are well-recognized. Vitamin A is another essential micronutrient throughout the development of heart. Its active form, retinoic acid, is involved in intracellular signaling, regulating development. Vitamin A has recently been found to play an antioxidant role in cardioprotection against ischemic heart disease as well. Low serum carotene has been linked with the increased risk of ischemic heart disease. Folic acid (vitamin B_9), a member of the B-complex family of vitamins, functions by S-methylation of homocysteine to generate methionine. Homocysteine, which is an intermediate formed during the metabolism of methionine, has recently been found to be an important risk factor for vascular disease including coronary atherosclerosis. Folic acid may be used to reduce increased homocysteine concentration in patients with coronary artery disease.

Several metal ions, especially divalent cations, are the important members of essential micronutrients. Copper is a catalytic cofactor for Cu/Zn-SOD and ceruloplasmin, both of which have been implicated in cardioprotection. Manganese is also a cofactor for Mn-SOD which plays an important role in cardioprotection. Targeted MnSOD gene delivery was found to ameliorate ischemia/reperfusion injury. MnSOD can also be induced in the heart during myocardial adaptation to ischemic stress. Glutathione peroxidase ($GSHP_x$) is an important member of the first line of antioxidant defense, which detoxifies hydrogen peroxide. Hearts of the transgenic mice overexpressing extra copies of $GSHP_{x-1}$ gene are resistant while knockout mice devoid of any copies of $GSHP_{x-1}$. are susceptible to ischemia reperfusion injury [7]. Selenium is a cofactor for the $GSHP_x$. The activity of $GSHP_x$. is decreased by selenium deficiency, which renders the myocardium susceptible to ischemic reperfusion injury. Selenium is also essential for humans because it protects the heart against cardiomyopathy. Two metal ions, zinc and magnesium, also play an important role in cardioprotection. Antioxidant effect of magnesium was found to play a role in its cardioprotective ability against ischemic reperfusion injury.

Recently, lycopene, the pigment responsible for the characteristic deep-red color of a ripe tomato, has been added to ever-growing essential micronutrients list because this pigment can quench singlet oxygen twice as efficiently as β-carotene. Dietary supplementation of lycopene significantly increased serum lycopene levels and lowered lipid peroxidation and LDL oxidation, two important risk factors of coronary heart disease [1]. Coenzyme Q_{10} is another important factor, which can provide cardioprotection against ischemia reperfusion injury. A recent study showed that supplementation of CoQ_{10} in the diet leads to the induction of ubiquitin gene expression simultaneously ameliorating ischemic injury to the heart [8].

Antioxidants and Redox Signaling

Despite the overwhelming evidence indicating reactive oxygen species as detrimental factors, it is becoming increasingly clear that these free radicals can also function as signaling molecules and hence, may be essential for the biological and physiological functions of the cells [3]. Biological cells including cardiomyocytes contain enzymes that can simultaneously generate reactive oxygen species and intracellular redox buffer in response to a specific stress. Depending on the amounts of antioxidant reserve and oxygen free radicals, the reactive oxygen species are either destroyed or survive. Thus, the oxygen free radicals fulfill the definition of a second messenger, which are either up-regulated or down-regulated after physiological stimuli like ischemia. A number of growth factors and cytokines have been found to induce oxidative stress and overproduce specific antioxidant enzymes. Environmental stresses including heat stress; oxidative stress and ischemia/reperfusion also produce oxidative stress, which is then translated into the induction of antioxidant enzymes.

Perhaps the finding that the production of reactive oxygen species during the agonist-induced activation of the nuclear transcription factor, NFκB, provided the first concrete evidence for the role of reactive oxygen species as a second messenger. NFκB regulates the inducible expression of a number of genes involved in cell survival and execution. For example, NFκB has been found to control anti-apoptotic gene, bcl-2, and pro-apoptotic factors, bax and p53, in the ischemic/reperfused myocardium [4]. Diverse extracellular signals from IL-1, TNFa, H_2O_2 etc. converge into development of oxidative stress, which leads to activation of NFκB. Such activation of NFκB can be blocked by antioxidants such as vitamin E or α-lipoic acid. Antioxidants such as N-acetyl cystein can prevent NFκB activation by diverse stimuli, including H_2O_2 by suppressing the generation of reactive oxygen species.

Although ischemia/reperfusion has been found to manipulate cardiomyocyte life and death through the redox signaling, precise physiological role of the signaling process is not clear. Ischemia/reperfusion, especially intermittent ischemia commonly known as ischemic preconditioning or adaptation, leads to the activation of both G-proteins and receptor tyrosine kinases potentiating a signaling cascade that results in the activation of multiple kinases. These sequential events lead to the induction of the activation of several redox-sensing transcription factors and genes [4]. Such intracellular events ultimately dictate whether the cells survive or die. The changes in gene expression are likely to influence the physiological function of the cardiomyocytes during post-ischemic survival.

A number of recent studies clearly demonstrated that redox signaling plays a physiological role in myocardial survival during post-ischemic period. As mentioned earlier, oxygen-free radicals are generated during ischemia/reperfusion. When heart is adapted to ischemic stress by repeated short-term ischemia and reperfusion, the generation of the reactive oxygen species is rapidly increased, but does not increase at the same rate during subsequent ischemia and reperfusion. The same pattern of the development of oxidative stress is observed when hearts are treated with endotoxin, IL-1 or TNFα. Interestingly, these interventions lead to the development of oxidative stress within a very short period, but reduce/inhibit subsequent oxidative stress development when hearts are subjected to ischemia/reperfusion. Additionally, such adaptive response is associated with the induction of the expression of a number of stress proteins, including HSP 27, HSP 70 and heme oxygenase and antioxidant enzymes that include Mn-SOD and glutathione peroxidase. Such adaptive response is also associated with the increased binding activity of NFκB. Increased activity of NFκB and induction of the protective proteins can be blocked by pretreating the hearts with an oxygen free radical scavenger such as DMTU [5]. More interestingly, an endotoxin-derived compound, monophosphoryl lipid A (MLA), was found to induce iNOS mRNA through a tyrosine kinase-dependent mechanism and protect the heart from ischemic reperfusion injury.

Thus, it seems likely that one of the major functions of redox signaling in the ischemic myocardium is to synthesize stress-inducible proteins comprising the *third line of defense* through the activation of transcription factors such as NFκB or by potentiating induction of other inducible proteins such as iNOS. However, further study is necessary to completely unveil the physiological function of redox cycling.

Summary and Conclusion

It should be clear from the above discussion that the potential for antioxidant therapy in cardiovascular diseases is enormous. It is a fact that the primary defense of heart against cellular injury comprises antioxidants. These antioxidants are lowered in the pathophysiological conditions in concert with massive production of reactive oxygen species warrants administration of exogenous antioxidants as potential therapy. Such antioxidant therapy may be manifested by nutritional supplement of the antioxidants or by direct administration into the blood circulation. Many of these antioxidants are found as polyphenolic compounds in fruits and vegetables. Regular consumption of fresh fruits and vegetables is expected to upregulate body's defense system against cardiovascular diseases. Antioxidant gene therapy has also great potential and is gaining popularity.

It is also important to remember that the same detrimental and cytotoxic reactive oxygen species can function as second messengers and become responsible for redox signaling leading to the induction of the expression of many cardioprotective genes. These genes which include anti-death gene Bcl-2, transmit survival signal and tend to protect the ischemic myocardium from the cellular injury especially, from the apoptotic cell death. Antioxidants by eliminating the oxidative stress may simply block this important survival signal.

References

1. Agarwal, S. and A.V. Rao (1998) Tomato Lycopene and low density lipoprotein oxidation: A human dietary intervention study. *Lipids* 33: 981–984.
2. Das, D.K., N. Maulik and I.I. Moraru (1995) Gene expression in acute myocardial stress. Induction by hypoxia, ischemia, reperfusion, hyperthermia and oxidative stress. *J. Mol. Cell Cardiol* 27: 181–193.
3. Das, D.K., N. Maulik, M. Sato and P. Ray (1999) Reactive oxygen species function as second messenger during ischemic preconditioning of heart. *Mol. Cell Biol.* 196: 59–67.
4. Maulik, N., S. Goswami, N. Galang and D.K. Das (1999) Differential regulation of Bcl-2, AP-1 and NFkB on cardiomyocyte apoptosis during myocardial ischemic stress adaptation. *FEBS Lett* 443: 331–336.

5. Maulik N., H. Sasaki, S. Addya. and D.K. Das (2000) Regulation of cardiomyocyte apoptosis by redox-sensitive transcription factors. FEBS Lett. 485: 7–12.

6. Maulik, N., T. Yoshida and D.K. Das (1998) Oxidative stress developed during the reperfusion of ischemic myocardium induces apoptosis. *Free Rad. Biol. Med.* 24: 869–875.

7. Maulik, N., T. Yoshida and D.K. Das (1999) Regulation of cardiomyocyte apoptosis in ischemic reperfused mouse heart by glutathione peroxidase. *Mol. Cell Biochem.* 196: 13–21.

8. Maulik, N., T. Yoshida, R.M. Engelman, D. Bagchi, H. Otani and D.K. Das Dietary coenzyme Q_{10} supplementation renders swine hearts resistant to ischemia reperfusion injury. *Am. J. Physiol.*

9. Otani, H., R.M. Engelman, J.A. Rousou, R.H. Breyer and D.K. Das (1986) Enhanced prostaglandin synthesis due to phospholipase breakdown in ischemic-reperfused myocardium. Control of its production by a phospholipase inhibitor or free radical scavengers. J. Mol Cell Cardiol. 18, 953–961.

10. Todd, S., H. Tunstall-Pedoe, M. Woodward and C. Bolton-Smith (1999). Dietary antioxidant vitamins and fiber in the etiology of cardiovascular disease and all causes mortality: results from the Scottish Heart Health Study. Am. J. Epidemiol. 150: 1073–80.

11. Tosaki, A., D. Bagchi, A. Hellegouarch, T. Palli, G.A. Cordis and D.K. Dass (1993) Comparisons of ESR and HPLC methods for the detection of hydroxyl radicals in ischemic/reperfused hearts. A relationship between the genesis of oxygen-free radicals and reperfusion-induced arrhythmias. *Biochem. Pharmacol.* 45: 961–969.

12. Maulik, N., H. Sasaki, S. Addya, and D.K. Das (2000) Regulation of Cardiomyocyte apoptosis by redox-sensitive transcription factors. *FEBS Letters* 485: 7–12.

Radiobiology and Bio-Medical Research
Edited by K.P. Mishra
Copyright © 2004 Narosa Publishing House, New Delhi, India

2. Reactive Oxygen Species in Lung Cancer

K.L. Khanduja, Rashmi Sharma and D. Behera

Departments of Biophysics and Pulmonary Medicine[#], Postgraduate Institute of
Medical Education & Research, Chandigarh-160 012, India

Abstract   Human alveolar macrophages (AMs) are the major source of
reactive oxygen species (ROS) involved in pulmonary injury. The antioxidant
defense system which controls the excessive generation of ROS is essential
in maintaining the tissue homeostasis. This study has revealed increased
ROS (superoxide, H_2O_2, total free radicals) in AMs retrieved from malignant
lobe of the lung cancer patients. Total glutathione, with a significant higher
concentration of oxidized glutathione, was significantly increased in the
broncho-alveolar lavage fluid aspirated from the malignant lobe of the cancer
patients. Whereas, HMP shunt activity and levels of vitamin A and vitamin
E were significantly decreased in AMs belonging to the malignant lobe as
compared to the AMs belonging to disease free lobe and lobes of patients
suffering from non-malignant lung diseases.

Introduction

Lung cancer is presently the most common malignancy in the world. Lungs are
unique as they have epithelial surface area that is at risk of oxidant mediated
attack both by the environmental factors, including cigarette smoke [11], and
activated phagocytes in particular alveolar macrophages (AMs). Macrophages
are capable of secreting a variety of cytokines such as tumor necrosis factors,
interleukins (IL-1, IL2) and interferon (IFN). Human AMs are the major source
of reactive oxygen species (ROS) such as superoxide (O2–), hydrogen peroxide
(H_2O_2), and possibly other free radicals which are involved in pulmonary injury.
Detoxification of reactive oxygen species is, therefore, one of the prerequisites
of aerobic life. The antioxidant mechanisms which control the excessive generation
of ROS thus, present a major line of defense regulating general health status.
Intracellular superoxide radicals are detoxified by cytoplasmic Cu, Zn, and SOD
to form H_2O_2 [7], which is further detoxified to H_2O by catalase and glutathione
peroxidase (GPx). Glutathione (GSH) is the most abundant nonprotein thiol in
the alveolar fluid of the lung which is able to scavenge H_2O_2 [20]. At the expense
of NADPH, GSH can be regenerated from oxidized glutathione. Oxidative stress
has been shown to stimulate the activity of hexose monophosphate (HMP) shunt
pathway, the main source of NADPH [16].

Reactive oxygen species are also scavenged nonenzymatically by the endogenous
antioxidants such as retinol (vitamin A) and α-tocopherol (vitamin E). Potent
anti-carcinogenic effects of retinoids have been demonstrated in a number of

studies [4, 12]. Antioxidant properties of α-tocopherol reflect its ability to neutralize free radicals including toxic oxygen intermediates, thereby preventing peroxidation of unsaturated lipids.

Studies on oxidant-mediated respiratory injury is of interest not only because it represents a major pathogenic mechanism of many lung diseases but also acts as a generic model to study oxidant-antioxidant imbalance in human disease. Since the human alveolar fluid (with its inflammatory cells and extracellular components, including antioxidants) is readily accessible by broncho-alveolar lavage (BAL), it is possible to gain access to biological components intimately related to lung injury. For the present work the alveolar macrophages were collected from malignant and disease-free lobes of the same individual separately for their comparison with macrophages from the patients suffering from other nonmalignant lung diseases.

Methods

One hundred and thirty five histologically proven lung cancer patients were enrolled at Nehru Hospital, Postgraduate Institute of Medical Education and Research, Chandigarh, India. For comparison, ninety subjects with non-malignant lung diseases were also included in the study. The non-malignant lung diseases were: interstitial lung disease, sarcoidosis, tuberculosis, chronic obstructive pulmonary disease and emphysema. Subjects with lower respiratory tract infection in the preceeding four weeks were excluded from the study. Similarly, those who had received specific therapy for lung malignancy, in the form of chemotherapy and/or radiotherapy, were also excluded from the study. All the subjects gave their informed consent for the procedure and the study fulfilled the criteria of the Ethical Committee of our institute.

Macrophages separated from bronchalveolar lavage were suspended in Earle's balanced salt solution and incubated at 37°C in 5 percent CO_2 and 95 percent air for an hour. The nonadherent cells were removed by washing. The nature of adherent cells was evaluated with Wright-Giemsa stain. In all cases the resulting adherent cells were > 92 percent alveolar macrophages. The cell viability was assessed by exclusion of trypan blue.

The procedure used to assess superoxide anion production by AMs was a modification of that originally described by Park *et al.* [16]. The amount of H_2O_2 released by AMs with and without PMA stimulation was estimated by the horseradish peroxidase method [17]. Total free radical activity in terms of chemiluminescence was measured in Micro β-plus Liquid Scintillation Counter, (Wallac) by the method of Allen and Loose [1]. AMs were washed three times with sterile phsophate buffered saline (PBS), at pH 7.4. The cells were lysed by three freeze thaw cycles followed by sonication for one minute. The cellular debris was then removed by centrifugation. Superoxide dismutase activity was measured in an aliquot of the supernatant according to the method followed by Kono [9]. Catalase activity was measured in the supernatant by the method followed by Luck [10]. Glutathione peroxidase was determined by following the procedure laid out by Paglia and Valentine [15].

Total glutathione (GSH+GSSG) was measured in BAL fluid using the assay of Sies and Akerboom [23]. Assay of GSH was performed by the method of Moron *et al.* [13].

HMP shunt pathway was determined as the amount of $^{14}CO_2$ liberated from sample incubated with D-$[1-^{14}C]$ glucose by the method followed by of Przybytkowski and Averill-Bates [18].

Vitamin A and E were measured in AMs by high performance liquid chromatography [3].

All data were expressed as the mean ± SD. Comparison between malignant and disease free lobe of lung cancer patients were analyzed with a paired t-test, whereas comparison with non-malignant lung diseases was analyzed with an unpaired student's test.

Results and Discussion

There was a significant increase in superoxide and hydrogen peroxide production by AMs from malignant lobe in comparison to non-malignant and disease-free lobes (Table 1).

Table 1. Levels of superoxide, hydrogen peroxide, superoxide dismutase, catalase, glutathione peroxidase and glutathione in alveolar macrophages of disease-free, non-malignant and malignant lung lobes of patients suffering from benign and malignant lung diseases. Values are mean ± SD

Parameter		Disease-free	Non-malignant	Malignant
O_2^-				
	Basal	1.60 ± 0.50	1.80 ± 1.07	5.06 ± 1.70[a,b]
	Stimulated	5.73 ± 1.09	6.20 ± 1.03	18.0 ± 2.23[a,b]
H_2O_2		18.3 ± 1.22	27.2 ± 3.29[a]	37.5 ± 3.94[a,b]
Superoxide dismutase		4.81 ± 0.64	5.03 ± 1.63	8.66 ± 2.09[a,b]
Catalase		453 ± 77.2	226 ± 49.2[a]	391 ± 61.4
Gluatathione peroxidase		56.2 ± 9.12	37.5 ± 7.99[a]	28.6 ± 4.68[a,b]
Gluatathione:				
	Reduced	35.2 ± 3.68	33.8 ± 8.47	19.1 ± 4.73[a,b]
	Oxidized	24.6 ± 6.92	20.0 ± 13.0	60.4 ± 13.4[a,b]
	Total	59.8 ± 6.38	53.8 ± 7.88	79.6 ± 10.8[a,b]

O_2^-, Formazan(+) cells/250 cells; H_2O_2, nmoles/30min/mg protein; SOD, IU/mg protein; catalase, nmoles of H_2O_2 decomposed/mg protein; Glutathione peroxidase, nmoles NADPH consumed/min/mg protein; Glutathione, nmoles/ml of BAL fluid.
a: Statistically significant w.r.t. disease-free.
b: Statistically significant w.r.t. non-malignant.

Lucigenin dependent chemiluminescence response of AMs in three groups is shown in Table 2. Basal as well as latex-induced chemiluminescence values were significantly ($p < 0.001$) higher in AMs of malignant lobe.

SOD activity was found to be significantly increased in AMs obtained from malignant lobe of lung cancer patients, whereas catalase and glutathione peroxidase activity of AMs of malignant lobe were found to be significantly decreased.

Table 2. Levels of vitamin A, vitamin E and total free radicals (Lucigenin-mediated chemiluminescence) and activity of hexose monophosphate (HMP) in alveolar macrophages of disease-free, non-malignant and malignant lung lobes of patients suffering from benign and malignant lung diseases. Values are mean $\pm$ SD

Parameter		Disease-free	Non-malignant	Malignant
Vitamin A		1.45 ± 0.20	4.47 ± 0.36^a	$0.47 \pm 0.08^{a,b}$
Vitamin E		0.66 ± 0.13	1.52 ± 0.29^a	$0.30 \pm 0.08^{a,b}$
Total free radicals				
	Basal	6.90 ± 1.07	4.98 ± 0.12	$10.2 \pm 1.25^{a,b}$
	Stimulated	10.2 ± 2.63	19.3 ± 4.36^a	17.2 ± 8.61^a
HMP				
	Basal	4.93 ± 2.32	3.97 ± 2.80	4.36 ± 1.9
	Stimulated	9.47 ± 5.32	5.31 ± 3.36	6.02 ± 2.48

Vitamin A and Vitamin E, $\mu g/10^6$ cells; Total free radicals, LCPS/10^6 cells; HMP, nmoles $^{14}CO_2$ released/30 min/10^6 cells.
(a) Statistically significant w.r.t. disease free.
(b) Statistically significant w.r.t. non-malignant.

Total glutathione concentration was significantly increased in the BAL fluid obtained from malignant lobe of lung cancer patients (Table 1). In addition, a significantly higher percentage of total glutathione was in the oxidized form (GSSG).

Rate of release of $^{14}CO_2$ as in index of overall HMP shunt activity of AMs from malignant lobe of cancer patients was found to be significantly decreased as compared to disease-free lobe. Similarly, vitamin A and E contents were significantly decreased in AMs retrieved from malignant lobe of lung cancer patients as compared to diseases-free lobe and non-malignant lung diseases.

Alveolar macrophages represent the main cellular component of defense system bringing about the protection of an organ and regulating cell and humoral response in the lungs [8]. In view of these observations, the present study assessed the function of macrophages from disease and disease-free lobes of the same subject to evaluate if these cells behave differentially within the same system. The increased levels of ROS in the macrophages of malignant lobe seem to promote the tumorigenesis. This hypothesis is supported by the results obtained by Slaga *et al.* [24] which provides the strong evidence that oxygen free radicals play a significant role in the promotion of tumors. The ROS generated by inflammatory phagocytes may cause injury to target cells which contribute to cancer development.

In the present study, 80 percent of patients were smokers, which was not surprising as tobacco smoking continues to be the most important etiological factor for lung cancer in Indian population. Numerous studies, including the present one, have demonstrated that *in vitro* treatment of macrophages [21] and neutrophils [22] with tumor promoters results in the production of superoxide anion radicals and hydrogen peroxide. The mechanisms of *in vivo* activation are still ill defined, but mediators such IL-1, gamma-IFN and tumor necrosis factor are likely to be implicated.

Increased attention has been paid to the question whether individual inherent status of reactive oxygen species can be associated with total free radical activity as a whole. Hence further studies to shed some light on the status of total free radical activity were carried out. Luminol could not enhance chemiluminescence in AMs. Monocytes transforming to macrophages lose their myeloperoxidase activity [2]. For this reason, it might be expected that the production of luminol dependent chemiluminescence by alveolar macrophages would be unlikely. In contrast, lucigenin does not require peroxidase activity for the production of chemiluminescence [5]. In view of this, the method was used to assess the generation for ROS as a whole. The total free radical activity in terms of enhanced chemiluminescence response cf AMs obtained from malignant lobe of lung cancer was found to be significantly increased which supports the observation of increased O_2^- levels and also suggests that oxidative burden was maximum in lower respiratory tract of patients with lung cancer.

The aforementioned assault of endogenous and exogenous oxidant mandates that all metabolically active cells assemble an antioxidant strategy that utilizes a broad array of resources to prevent or limit oxidant injury. Glutathione is an important non-protein thiol and is a powerful nucleophile which can participate in a variety of enzyme link detoxification reactions. The lung has substantial transpeptidase activity and it utilizes extracellular GSH that happens to be available as a component of alveolar fluid. Certainly, in patients subjected to an increased oxidant stress such as patients of lung cancer, one might expect a preponderance of oxidized glutathione. The current study demonstrates a significant portion (76 percent) of glutathione in the oxidized form (GSSG) which may indicate increased oxidative stress leading to greater demand to combat the free radicals in the lower respiratory tract of lung cancer patients. The supply of GSH in the cells is maintained by the reducing equivalents which are supplied through the hexose monophosphate shunt pathway. Thus, the NADPH-regenerating system can become pivotal in antioxidant defense mechanism for cells and prevent the depletion of thiols. The formation of $^{14}CO_2$ was taken as an index for the HMP shunt activity. The HMP shunt activity of AMs from malignant lobe was significantly decreased. Glucose-6-phosphate dehydrogenase and 6-phosphogluconate dehydrogenase in the hexose monophosphate shunt are important antioxidant enzymes by virtue of reactions catalyzed by them to supply NADPH. Excessive formation of ROS in lung diseases might be the critical factor for inactivation of these enzymes.

Besides glutathione, enzymatic antioxidants such as superoxide dismutase, catalase, glutathione peroxidase and nonenzymatic antioxidants like vitamin A and E are equally important in combating the ROS. Activity of SOD was significantly increased. The mechanism of the induction of antioxidant enzyme-SOD in AMs is not known. However, the induction of antioxidant gene is presumed to be involved [19]. In contrast, a decreased activity of catalase and GPx was observed. GPx is a selenium containing enzyme that inactivates both hydrogen and lipid peroxides. The decreased GPx levels may reflect selenium defect in patients with lung cancer. Moreover, this may be due to the action of highly electrophilic free radicals on the active site of the enzyme. The enhanced formation of superoxide radical [19] followed by increased activity of SOD would lead to increased formation of hydrogen peroxide due to dismutation of superoxide radical. The impaired detoxification of H_2O_2 due to depletion of catalase or GPx may lead to enhanced levels of H_2O_2 which is a precursor for formation of hydroxyl radicals involved in cancer initiation as well as in cancer promotion.

Vitamin A and E are antioxidant vitamins that show protective effects against oxidative stress. These micronutrients have been found as late stage inhibitors of human cancer. It was observed that retinol and α-tocopherol levels were decreased in AMs from malignant lobe of lung cancer patients. The low levels of antioxidant vitamins could be possibly due to greater utilization of vitamin A and E by rapidly growing cells of the tumor site. This may also be due to greater conversion of retinol to its inactive metabolite that is anhydroretinol [14] and oxidation of vitamin E to quinone metabolite. Moreover, as the malignant lobe is chronically exposed to a greater assault of oxidants [21], it is conceivable that these low levels are the results of an increased demand for these antioxidants or enhanced degradation of these micronutrients ROS.

To summarize, increased oxidant burden in terms of constitutive generation of ROS by AMs from malignant lobe is accompanied by decrease in antioxidant protection. This oxidant antioxidant imbalance in the AMs from malignant lobe could potentially enhance the neoplastic behavior by augmenting both genetic instability of the tumor and its capacity to injure and penetrate the host tissue. The enhanced levels of ROS in AMs from the tumor bearing lobe may further aggreviate the tumor leading to massive progression.

References

1. Allen, R.C. and L.D. Loose (1976) Phagocytic activation or f luminol dependent chemiluminescence in rabbit alveolar and peritoneal macrophages Biochem. Biophys. *Res. Coomun.* 69: 245–252.
2. Breton Gorius, J., J. Guichard, W. Vainchencker and J.L. Vilde (1980) Ultrastructural and cytochemical changes induced by short and prolonged culture of human monocytes. *J. Retico. Soc.* 27: 289–301.
3. Catigani, G.L. and J.G. Bieri (1983) Simultaneous determination of retinol and *a*-tocopherol in serum or plasma by liquid chromatography. *Clin. Chem.* 29(4): 708–712.
4. Chytil, F. (1992) The lungs and vitamin A. Am. J. Physiol. 262: L517–L527.
5. Gyllenhammar, H. (1987) Lucigenin chemiluniscence in the assessment of neutrophil superoxide production. *J. Immunol Methods* 97: 209–213.
6. Halliwell, B. and J.M.C. Gutteridge (1989) Free radicals in Biology and Medicine second edition. Oxford, England: Clarendon Press.
7. Heffner, T.M., L.A. Brown and D.P. Jones (1989) Pulmonary strategies of antioxidant defense. *Am. Rev. Respir Dis.* 140: 431–554.
8. Iakovlev, M.L., L.D. Zubairova, A.N. Krupnik and N.K. Permiakov (1991) Alveolar macrophages of lung physiology and pathology. *Arkh. Patholo.* 53(4): 3–8.
9. Kono, Y. (1978) Generation of superoxide radical during auto-oxidation of hydroxylamine and an assay of superoxide dismutase. Arch. Biochem. Biophys. 186: 189–195.
10. Luck, H. Catalase: In Bergmeyer H_2O_2 ed, Methods of enzymatic analysis. New York: Academic Press. Section 3: 885–894.

11. Ludwig, P.W., B.A. Schwartz, J.R. Hoidal and D.E. Niewoeher (1985) Cigarette smoking caused accumulation of polymorphonuclear leukocytes in lung parenchyma. *Am. Rev. Respir. Dis.* 131: 828–830.

12. Moon, R.C., R.G. Mehta and K.V.N. Rao (1994) Retinoids and cancer in experimental animals. In M.G. Spron, A.B. Roberts and D.S. Goodman, editors. *The Retinoids: Biology, Chemistry and Medicine* second edition. 572–595, New York, Raven Press.

13. Moron, M.S., J.W. Dipierri and B. Mannervik (1979) Levels of glutathione, glutathione reducatase and glutathione S. transferase activities in rat lung and liver. *Biochem. Biophy. Acta.* 582: 67–78.

14. Muto, Y., and H. Moriwaski (1984) Antitumor activity of vitamin A and its derivatives. *J. Natl. Cancer Inst.* 73: 1389–1393.

15. Paglia, D.E. and W.N. Valentine (1967) Studies on the quantitative and qualitative characterization of erythrocyte glutathione peroxidase. *J. Lab. Cln. Med.* 70: 158–169.

16. Park, B., S. Fikrig and H. Smithwick (1968) Infection and nitroblue tetrazolium reduction by neutrophils. A diagnostic act. *Lancet* ii, 532–534.

17. Pick, E. and Y. Keisari (1980) A simple colorimetric method for the measurement of H_2O_2 produced by cell in culture. *J. Immunol Methods* 38: 161–170.

18. Przybytkowski, E. and D.A. Averill Bates (1996) Correlation between glutathione and stimulation of pentose phosphate cycle in situ in chinese hAMster ovary cells exposed to hydrogen peroxide Arch. *Biochem. Biophy.* 325 (1) 91–98.

19. Rahman, I. and W. Machnee (1996) Oxidant/antioxidant imbalance in smokers and chronic obstructive pulmonary disease. *Thorax* 51: 348–350.

20. Sagone, A.L., R.M. Husney, M.S. O'Dorisio and E.N. Metz (1984) Mechanism of oxidation of reduced glutathione by stimulated granulocytes. *Blood* 63: 96–104.

21. Sharma, R.N., D. Behera and K.L. Khanduja (1997) Increased reactive oxygen species production by alveolar macrophages from malignant lobe of lung cancer patients. *J. Clin Biochem. Nutr.* 22(3): 183–196.

22. Sharma, R.N., D. Behera and K.L. Khanduja (1997) Reactive oxygen species formation in peripheral blood neutrophils in different types of smokers. *Ind. J. Med. Res.* 106: 475–480.

23. Sies, H. and T.P.M. Akerboom (1989) Glutathione disulphide (GSSG) efflux from cells and tissues. *Methods Enzymol.* 105: 445–451.

24. Slaga, T.J., A.L.P. Klein Szanto, L.L. Triplett and J.E. Yotti Lpand Trosko (1981) Skin tumor promoting activity of benzoyl peroxide a widely used free radical generating compound Science, 213: 1023–1025.

Radiobiology and Bio-Medical Research
Edited by K.P. Mishra
Copyright © 2004 Narosa Publishing House, New Delhi, India

3. Calcineurin as a Putative Biomarker of Oxidative Stress and Disease

C. Subramanyam

Department of Biochemistry, Osmania University, Hyderabad 500 007, India

Introduction

Reactive oxygen species (ROS), generated during oxidative metabolism and energy production in the body, are involved in diverse biological functions including enzyme catalyzed reactions, electron transport in mitochondria, activation of nuclear transcription factors, antimicrobial action of neutrophils and macrophages as well as in aging. Along with reactive nitrogen species, they are involved in normal cell regulation wherein oxidants and redox status are important facets of signal transduction. A delicate balance exists between the generation of ROS and antioxidants in health. Initiation of disease is usually associated with an imbalance between them, caused by excessive generation of ROS resulting in oxidative stress. ROS generated at relatively high levels by activated leukocytes have been implicated in cell damage through oxidative modification of cellular macromolecules. While oxidative stress could be the primary cause of diseases such as cancer, autoimmune disease, renal failure, neurodegenerative diseases and several others, it could also be involved as a secondary complication in diseases such as diabetes mellitus leading to vascular and other complications.

Identification of specific biomarkers indicative of oxidative stress would thus provide a useful means in the diagnosis and treatment of at least some of the above mentioned diseases. H_2O_2 is an important ROS formed *in vivo* both by dismutation of superoxide and upon catalysis mediated by a range of oxidases metabolizing monoamines, xanthine, urate and D-amino acids. The clinical relevance of elevated H_2O_2 levels in breath (in respiratory disorders), blood, tumor cells and tissues (as in myocardial dysfunction) is well recognized. Recent studies have also demonstrated the excretion of H_2O_2 in human urine samples, though its clinical relevance in disease conditions is still to be evaluated.

One of the intracellular targets for H_2O_2 is calcineurin, the Ca^{2+}/calmodulin dependent neutral phosphatase. In addition, the importance of calcineurin in immune systems (that are quite often affected in disease) emanates from its role in activation and proliferation of T lymphocytes and its susceptibility to immunosuppressive drugs such as cyclosporin and FK 506. The activity of this metalloenzyme is subject to redox regulation in view of the bimetallic center $(Fe^{3+} - Zn^{2+})$ in its active site. ROS, including H_2O_2 can thus inhibit its activity and superoxide dismutase has been shown to protect the enzyme from such damage. Since these two molecules share certain similarities with regards to their wide distribution in body fluids, it seems reasonable to assume that clinical evaluation of both calcineurin and H_2O_2 may yield information on oxidative stress under conditions of health and disease.

Even though calcineurin has so far been recognized as an intracellular enzyme, we have recently reported (a) its extracellular presence in biological fluids such as serum and amniotic fluid and (b) evolved simple methods to quantify its activity and content in such extracellular fluids [1]. In continuation of these findings and the forementioned hypothesis, studies were conducted in our laboratory to evaluate calcineurin as a biomarker of oxidative stress-related disease.

Calcineurin in the Etiology of Lymphopenia Associated with Chronic Renal Failure

A number of studies have implicated the increased activity of ROS in chronic renal failure (CRF), as evidenced by increased malondialdehyde (MDA) levels in serum and erythrocytes. Our earlier studies have also demonstrated decreased activity of superoxide dismutase (SOD) in sera, associated with progressive oxidation of low-density lipoproteins in CRF [2]. Interestingly, it has earlier been reported that SOD protects calcineurin – a Ca^{2+}-calmodulin dependent serine/threonine protein phosphatase, from ROS mediated inactivation [4]. Importantly, renal transplant patients could be segregated into populations with different sensitivities to cyclosporine inhibition based on calcineurin activity of peripheral blood mononuclear cells. Since calcineurin is now known to be involved in T-lymphocyte proliferation, this raises the possibility that interplay between ROS and calcineurin may also be involved in the etiology of lymphopenia observed in CRF.

Studies were thus conducted in patients diagnosed for chronic renal failure and categorized, based on their serum creatinine levels. Results obtained in this regard indicated a marked decrease (75 percent) in serum calcineurin activity occurred even in stage I although further decrease occurred reaching a maximum of 88 percent in stage III. In comparison, the decreases in calcineurin activity of lymphocytes were much more gradual. The activities were 68, 40 and 17 percent of the control activities during stages I, II and III of CRF respectively. Such a gradual decrease in lymphocyte calcineurin activity may be due to increased activity of ROS caused by decreased SOD activity in serum. This would explain the gradual lymphopenia observed during CRF progression. The factors responsible for decrease in serum calcineurin activity are not properly understood although it is possible that it might be related to calcineurin activity of lymphocytes. In conclusion, a possible mechanism to account for gradual lymphopenia that occurs during the CRF could be suggested. Impairment in T cell activation caused by oxidative stress and decreased lymphocyte calcineurin activity seems to be involved in the lymphopenia encountered in CRF.

Calcineurin and Diabetes Mellitus

Enhanced production of ROS is now established to contribute to complications in insulin dependent and non-insulin dependent varieties of diabetes mellitus. Recent research has also focused on possible interactions between oxidative stress and cell signaling molecules in the development of diabetic complications. It has been suggested that redox-sensitive, molecular factors such as IRS-1, PI-3 and NF KB can become pathogenic factors due to imbalances in oxidative stress experienced in diabetes. In addition, decreased levels of serum calmodulin (the ubiquitous calcium binding protein) under conditions of experimental diabetes suggest a comprehensive understanding of free radical physiology in diabetes may be obtained by employing calcineurin as a biomarker for the oxidative stress encountered in diabetes. In continuation of these, a study was conducted in diabetic patients categorized into three groups based on their serum MDA levels. The observed increases in serum MDA in diabetes indicate a progressive decrease in antioxidant functions and other protective mechanisms in the patients. Simultaneously it was noted that serum calcineurin activity is also decreased in the patients and a positive correlation could be obtained between these two parameters studied. Such a finding suggests that serum calcineurin assay may be useful in assessing the oxidative damage under conditions of diabetes.

Calcineurin and Calmodulin in Leukemic Disease

Ca^{2+}-calmodulin dependent protein kinases and protein phosphatase (calcineurin) are known to play a significant role in growth and differentiation of lymphocytes. The involvement of calmodulin binding proteins in mediating several Ca^{2+}-calmodulin related events is also recognized [3]. Quantitation of calmodulin (CaM) levels in lymphocytes obtained from patients as well as in acute lymphoblastic leukemia cell line has earlier revealed an increase in its content. Further, increased expression of calcineurin could be induced during inhibition of HL-60 cell proliferation by all trans retinoic acid (ATRA) as well as 1, 25 α dihydroxy vit D_3. Identification of a protein immunologically similar to calcineurin in adult sera (Padma and Subramanyam, 1999) and reports on the involvement of CaN in lymphocyte proliferation provide the basis for its quantitation in hematopoietic malignancies such as the leukemias. In view of these reports, it was proposed to quantitate the content of CaM and CaN in the sera of patients diagnosed for the different types of leukemias with intent of ascertaining its application in the diagnosis and or prognosis. Thus, the contents of calmodulin and calcineurin and the activity of calcineurin were determined in the sera of patients diagnosed for acute lymphoid leukemia (ALL), acute myeloid leukemia (AML) and chronic myeloid leukemia (CML). Assaying calcineurin activity in leukemic cell lines also validated results obtained from such studies. In addition, assaying serum calcineurin activities in patients prior to and after treatment (with conventional therapeutic approaches) further established the prognostic application of serum calcineurin in leukemia treatment.

A salient observation made during this study was the decrease in the activity of calcineurin in the sera of patients diagnosed for different varieties of leukemias. This observation is in tune with earlier reports on the relatively low calcineurin activity in lymphocytes of patients diagnosed with leukemia. In addition, it was also noted that all varieties of leukemia are characterized by decreased serum calcineurin activity without associated changes in content of either calmodulin or calcineurin in serum. While acute lymphoid leukemia and chronic myeloid leukemia revealed 75 percent decrease in calcineurin activity, acute myeloid leukemia demonstrated an 87 percent decrease in the activity in comparison to the controls. Another important observation is the restoration of normal calcineurin activity following treatment in acute myeloid leukemia. However, such an effect could not be observed in acute lymphoid and chronic myeloid leukemias. Although significant differences were noted with respect to the activity, similar changes could not be observed with respect to either the content of calcineurin or calmodulin.

The results pertaining to changes in calcineurin activity in sera of patients diagnosed with acute and chronic leukemias were further confirmed and validated by assaying the calcineurin activity in leukemia cell lines of lymphoid (Molt-4) and myeloid origin (HL-60 and K-562). The most significant finding was the decrease in the calcineurin activity of acute leukemia cell lines similar to that observed in the sera of patients diagnosed with acute leukemia. In contrast to the serum levels, the contents of calcineurin and calmodulin were increased in all the varieties of leukemic cell lines studied. Activity of calcineurin in Molt-4, an acute lymphoblastic leukemia cell line decreased by 97 percent in comparison to the activity in normal lymphocytes. HL-60, an acute myeloid leukemia cell line also demonstrated an 85 percent decrease in the activity of calcineurin while K-562, the chronic myeloid leukemia cell line demonstrated an increase in the activity of calcineurin.

The results depicted above assume significance in view of earlier reports on (i) inhibition of the phosphatase activity of calcineurin by oxidants such as H_2O_2, superoxide and glutathione disulfide in a dose dependent manner and (ii) oxidative stress induced apoptosis in the leukemia cell line, HL-60. Further, decrease in the calcineurin phosphatase activity in two T-cell leukemic cell lines (Warzburg and Jurkat cell lines) was associated with an increase in hydrogen peroxide production. These studies provide a basis to explain the decrease in calcineurin activity observed in two of the three leukemic cell lines. The increase in the content of calmodulin in all the leukemic cell lines studied is comparable to earlier reports on the increase in CaM contents in leukemic cells as well as leukemic cell lines.

In conclusion, it seems reasonable to recommend the assay of serum calcineurin activity in diseases related to oxidative stress. It is pertinent to point out the ease with which the assay can be performed routinely. Conducting the phosphatase assay in presence and in absence of trifluoperazine permits quantitation of calmodulin-dependent phosphatase activity in the samples. Even though this method is less sensitive than the more specific assays using labeled substrates specific to calcineurin, this relatively inexpensive, non-hazardous and rapid method would offer distinct advantages in routine clinical analysis in the evaluation of oxidative stress related disease.

References

1. Padma, S. and C. Subramanyam (1999) *Clinical Biochemistry.* 32: 491–494.
2. Sasikala, M., C. Subramanyam and B. Sadasinudu (1999) *Indian Journal of Clinical Biochemistry*, 14: 176–183.
3. Subramamyam, C., S.C. Honn, W.C. Reed and G.P.V. Reddy (1990) *J. Cell. Physiol.* 144: 423–428.
4. Wang, X., V.C. Culotta and C.B. Klee (1996) *Nature* 383: 434–437.

Radiobiology and Bio-Medical Research
Edited by K.P. Mishra

4. Antioxidants and Anticancer Treatment

Benny H.K. Tan

Department of Pharmacology, Faculty of Medicine,
National University of Singapore, Singapore

Some Statistics

In the USA, 1.2 million new cases of cancer are diagnosed annually. Each year, 600,000 die of the disease. Singapore, with a population of 3.5 million people as of 1997, had an average of 6,300 cases a year during the period 1993–97 [5]. Cancer contributed to more than a quarter (25.6 percent) of all deaths during this period. This compares to 20.9 percent during 1988–92 and 14.8 percent during 1968–72, showing an increase in the cancer death rate.

There is an increasing realization that the efficacy of standard anticancer therapy, namely. surgery, x-irradiation and chemotherapy has reached a plateau. It appears that new modalities of treatment are needed to improve the response to anticancer therapy.

The search continues for new ways to augment anticancer treatment. VandeCreek *et al.*, (1999) surveyed the use of alternative therapies among 113 breast cancer outpatients compared with the general population and found that the number of breast cancer patients who took vitamins was 23 percent more than that in the general population.

Why Do Cancer Patients Take More Vitamins?

Epidemiological studies seem to support the idea that achieving high blood concentrations of vitamin C [1] and other vitamins (E or beta-carotene) [2] or having a regular dietary intake of vitamins can protect against cancer. With the added influence of aggressive marketing of the antioxidant benefits of high vitamin intake, healthy people today are taking vitamin supplements regularly and, in some countries, doctors are using them as part of cancer treatment regimens.

Vitamins with Antioxidant Actions

The vitamins that have antioxidant actions are:
vitamin A *(retinol)* and its analogs
beta-carotene and polar carotenoids (PC)
vitamin C (mainly *sodium ascorbate*)
vitamin E (mainly *alpha-tocopherol succinate* [ATS]).

Vitamin A
Vitamin A (retinol) and its analogs, all-trans retinoic acid (RA) and 13-cis retinoic acid (isotretinoin), are the forms available. RA is formed in human tissues from retinol and beta-carotene. RA and its derivatives can induce cell differentiation and inhibit growth of cancer cells *in vitro* by increasing the levels of growth-inhibitory signals. These include inhibition of protein kinase C [15] and reduced expression of c-myc and H-ras oncogenes *in vitro* [25].

At high doses, vitamin A has been reported to have a relatively selective effect in that it reduces the growth of transplanted tumours in animals without affecting proliferating organ systems [29]. However the problem of extreme toxicity at high doses (headache, lethargy, anorexia, vomiting, visual disturbance), especially with some RA derivatives, reduces the possibility of their usefulness in cancer treatment.

Carotenoids
More than 1,000 carotenoids are found in nature; few occur in abundance in fruit and vegetables. The common examples are *beta-carotene* (carrot), *lycopene* (tomatoes, water melon) and *lutein* (spinach). Beta-carotene is available as the synthetic form and is the most widely studied carotenoid. 'Polar carotenoids' (PCs) are stable polar substances present in some synthetic beta-carotene preparations as well as natural sources of beta-carotene [8].

Concern with the safety of beta-carotene has arisen from a report on its association with a higher risk of lung cancer in smokers compared to the general population [10]. Aside from this, long-term intake of beta-carotene even at high doses has not been associated with toxic effects.

Some Evidence for the Anticancer Effects of Carotenoids
1. Synthetic beta-carotene caused marked regression of oral leukoplakia in humans [12].
2. Synthetic beta-carotene has been shown to increase the expression of the connexin43 gene, which encodes a major gap junction protein [35].

More human studies, however, are needed to support the notion that beta-carotene can be used in the treatment of human cancer.

Vitamin C
The common forms of vitamin C used in human and animal studies are ascorbic acid and sodium ascorbate. Vitamin C has been shown to:

(a) inhibit the growth of rodent and human tumor cells in a concentration-dependent manner *in vitro* [23]
(b) stimulate the growth of human parotid carcinoma cells but not adenoma cells at low doses *in vitro* [27] and also human leukemia cells *in vitro* [22]. The effects of vitamin C thus appear to be dependent on dose as well as cancer cell type.

Is Vitamin C Effective in Treating Tumours in Humans?

The results from human studies have been controversial and do not seem to support the use of vitamin C singly in anticancer treatment. Cameron *et al.*, (1979) reported that high doses, that is, 10g or more reduced the growth of several human tumors, increased chances of survival and provided better quality of life. However, Creagan *et al.* (1979) disputed the effectiveness of vitamin C in patients with advanced cancer. This discordant finding may be because of differences in the type of vitamin C used (ascorbic acid vs sodium ascorbate) and the form and stage of the tumors treated.

Unlike with vitamin A, there is no evidence of toxicity with vitamin C at high doses, except for occasional gastric pain.

Vitamin E

Prior to 1980, the form of vitamin E used in human and animal studies was either natural (alpha-tocopherol) or its acetate derivative. Because of their relative lack of stability in aqueous conditions, their effects on cell cultures could not be determined. In 1982, the succinate derivative, alpha-tocopherol succinate (ATS), was introduced and found to be the most active form of vitamin E. ATS appears to be the best choice among vitamin E derivatives for use in clinical studies.

The mechanism of action of ATS is similar to that of vitamin A. It inhibits *protein kinase C* activity [21] as well as expression of c-myc and H-ras oncogenes [24]. ATS also increases the synthesis and release of a growth inhibitory signal, transforming growth factor-beta [17], and inhibits the activation of E_2F [32], an important element in cell growth regulation.

B Vitamins

Most of the B vitamins have no direct anticancer activity. High doses of B_6 reportedly stimulated the growth of transplanted human lung cancer cells in nude mice. Such doses cause peripheral neuropathy in man. B_3 (nicotinamide) caused increased blood flow to tumor tissue in rats exposed to x-irradiation. The effects from using B vitamins individually may thus be counter-productive to cancer treatment.

Individual Antioxidant Vitamins in Combination with Standard Anticancer Agents

The question arises as to whether the concomitant use of antioxidant vitamins with standard anticancer modalities would enhance the overall anticancer effect. Vitamins A, C, ATS and beta-carotene have each been shown to enhance the growth inhibitory effects of standard therapy and of x-irradiation on cancer cells both *in vivo* and *in vitro*.

Table 1 [23] shows the effects of vitamin C in combination with chemotherapy or x-irradiation against murine neuroblastoma cells *in vitro*. Clearly, vitamin C enhanced the anticancer effects of 5-fluorouracil (5-FU), bleomycin and sodium butyrate as well as x-irradiation.

Table 1. Effect of vitamin C in combination with x-irradiation and certain chemotherapeutic agents on murine neuroblastoma cells in culture

Treatments	Cell Number (% of controls)
Sodium (Na) ascorbate (5 or 200 μg/ml)	105 ± 9
5-Fluorouracil (0.08 μg/ml)	62 ± 5
Na ascorbate (5 μg/ml) + 5-Fluorouracil	4.5 ± 2.1
X-irradiation (4 Gy)	28 ± 3
Na ascorbate (5 μg/ml) + X-irradiation	1.8 ± 0.4
Bleomycin (0.004 unit/ml)	27 ± 2
Na ascorbate (200 μg/ml) + Bleomycin	8 ± 1
Sodium butyrate (0.5 mM)	13 ± 2
Na ascorbate (200 μg/ml) + sodium butyrate	4.6 ± 1

from Prasad, 1980

Vitamin A or synthetic beta-carotene in combination with x-irradiation or cyclophosphamide increased the cure rate from 0 percent to 90 percent in rats with transplanted human breast adenocarcinoma [29], while vitamin E reportedly enhanced the effect of 5-FU against colon cancer cells [6].

Many other examples of such data directly demonstrate that the vitamins, rather than protecting cancer cells against free radical and growth-inhibitory effects of standard anticancer drugs, enhanced their anticancer effects and protected normal cells from their toxic effects.

A Mixture of Vitamins Can be Efficacious in Cancer Treatment

The use of single vitamins at very high doses appears to have no biological or clinical merit because toxicity arises when these are used at such doses. At lower doses, however, they may be ineffective or even stimulate cancer cell growth. The question arises whether individual vitamins can interact with each other to produce a higher degree of growth inhibition of cancer cells. If so, lower doses of multiple vitamins may then be used.

Prasad et al (1994) [see Table 2 below] reported that a mixture of four antioxidant vitamins viz. 13-cis RA, sodium ascorbate, ATS and polar carotenoids without beta-carotene produced a higher degree of growth inhibition of cultured human melanoma cells than single vitamins.

These observations indicated that individual vitamins do not affect cancer cell growth. However when used as part of a mixture with other vitamins, they appear to be able to inhibit cancer cell growth. Using a mixture of vitamins also avoids the risk of toxicity from using vitamins singly at higher doses.

Gey (1993) reported that for optimal synergistic protection, the plasma antioxidant levels should simultaneously exceed the threshold values of 28–30 umol/L lipid-standardized vitamin E, 40-50umol/L vitamin C, 0.4-0.5 umol/L carotene and 2.2-2.8 umol/L lipid-standardized vitamin A. It appears that from

Table 2. Effect of a mixture of four vitamins on growth of human melanoma cells in culture

Treatments	Cell number (% of controls)
Vit. C (100 μg/ml) + PC (10 μg/ml) + α-TS (10 μg/ml) +RA (7.5 μg/ml)	13 ±1
Vit. C (50 μg/ml) + PC (10 μg/ml) + α-TS (10 μg/ml) + RA (7.5 μg/ml)	56 ±3
Vit. C (50 μg/ml) + PC (5 μg/ml) + α-TS (5 μg/ml) + RA (3.8 μg/ml)	98 ±4
Vit. C (100 μg/ml)	64 ±3
Vit. C (50 μg/ml)	102 ±5
PC (10 μg/ml)	96 ±2
α-TS (10 μg/ml)	102 ±3
RA (7.5 μg/ml)	103 ±3

from Prasad *et al.* (1994)

these preliminary findings that properly designed clinical studies using multiple vitamins at appropriate doses need to be conducted to confirm the efficacy of vitamin mixtures in cancer treatment.

Conflicting Views on the Use of Antioxidant Vitamins in Cancer Therapy

Labriola and Livingston (1999) recommended that no supplementary antioxidants should be given concurrently with anticancer drugs which act by generating free radicals to cause cellular damage and necrosis of cancer cells. However, considering that epidemiologic studies to investigate the value of vitamins in the prevention of cancer have shown a strong link between diet and cancer, it appears that vitamins, because of their antioxidant and anticancer actions, warrant further evaluation as a therapeutic intervention in cancer patients.

Do Antioxidant Vitamins Reduce Toxicity of Standard Anticancer Therapy?

Another beneficial effect of vitamins in anticancer treatment that warrants consideration is the reduction in toxicity associated with standard anticancer therapy. Several animal studies have demonstrated that vitamins in combination with standard therapeutic agents may reduce the toxic effects of these agents.

Vitamin E was found to reduce lung fibrosis due to bleomycin [34] and cardiac toxicity induced by adriamycin [30]. Vitamin C has also been reported to protect normal cells against adriamycin toxicity [11]. Trizna et al. (1993) reported that vitamin C, ATS and RA reduced bleomycin-induced chromosomal breakage. More recently, vitamin A and beta-carotene were reported to reduce the adverse effects of cyclophosphamide and x-irradiation in mice [7].

Vitamin Supplementation in Anticancer Treatment

Physicians need to be aware of the body of evidences showing a positive effect of vitamins in the period following chemotherapy administration. Supplemental therapies have been shown to result in a higher percentage of successful outcomes when they are taken during this period. Prasad *et al.* (1999) have made the following recommendation for vitamin supplementation in cancer treatment:

1. Multiple antioxidant vitamins, including B vitamins
2. Additional vitamin C in the form of calcium ascorbate, up to 10g or more
3. Additional natural vitamin E 800 IU in the form of ATS
4. Additional beta-carotene 60 mg/day.

The vitamins are to be taken orally in two divided doses daily before meals. They are started 48 hours prior to standard therapy and continued until one month after completion of standard therapy. The doses of beta-carotene, vitamins C and E are then reduced gradually by one half over a four week period. The multivitamins together with the reduced doses of beta-carotene, vitamins C and E are continued for life.

Importance of Diet and Life-Style Modification to Vitamin Supplementation

In addition to vitamin supplementation, the diet should be low in fat (<10 percent of calories) and high in fibre (25-30g of fruits and vegetables). Life-style changes include abstinence from smoking, reduced consumption of caffeine, alcoholic beverages, daily exercise, reduced physical and mental stress.

Conclusion

Does supplementation with high doses of multiple antioxidant vitamins (together with dietary modification and lifestyle changes) improve the efficacy of standard anticancer therapy?

Random controlled clinical trials, the gold standard by which the clinical efficacy of any treatment is evaluated, are needed to test the hypothesis that vitamin supplements, diet and life-style changes can improve the efficacy of standard therapies by enhancing the survival and/or the quality of life of cancer patients.

Acknowledgment

The author thanks the National University of Singapore for providing financial support for his participation in the 88th Indian Science Congress, January 1–7, vary 2001 at New Delhi, India.

References

1. Block, G. (1991) Vitamin C. and cancer prevention: the epidemiologic evidence. *Am J. Clin. Nutr.* 53: 270S–282S.

2. Block, G., B. Patterson, A. Subar (1992) Fruit, vegetables and cancer prevention: a review of the epidemiologic evidence. *Nutr Cancer* 18 (1): 1–29.

3. Boik, J. (1996) *Cancer and Natural Medicine: A textbook of Basic Science and Clinical Research.* Oregon Medical Press.

4. Cameron, E., L. Pauling and L. Liebowitz (1979) Ascorbic acid and cancer. A review. *Cancer Research* 39: 663–681.

5. Chia, K.S., A. Seow, H.P. Lee and K. Shanmugaratnam, Cancer incidence in Singapore *Singapore Cancer Registry Report No. 5.*

6. Chinery, R., J.A. Brockman, M.O. Peeler, Y. Shyr, D. Beauchamp and R.J. Coffey (1997) Antioxidants enhance the cytotoxicity of chemotherapeutic agents in colorectal cancer: A p53-independent induction of $p^{21WAFI/CIPI}$ via C/EBPß. *Nature Medicine* 3: 1233–1241.

7. Cole, W.C. and K.N. Prasad (1997) Contrasting effects of vitamins as modulators of apoptosis in cancer cells and normal cells: a review. *Nutrition and Cancer* 29: 97–103.

8. Cole, W.C., K.N. Prasad (1998) Heterogeneity of commercial *b*-carotene preparations: correlation with biological activities. In Prasad K.N., Cole W.C. (eds) *Cancer and Nutrition.* Amsterdam:IOS Press. pp 99–105.

9. Creagan, E.T., C.G.O. Moertel and J.R. Fallon (1979) Failure of high dose vitamin C (ascorbic acid) therapy to benefit patients with advanced cancer. *N. Engl. J. Med.* 301: 687–690.

10. DeLuca, L.M. and S.A. Ross (1996) Beta-carotene increases lung cancer in cigarette smokers. *Nutr Review* 54: 178–180.

11. Fujita K., K. Shinpo, T. Sato, H. Niime, M. Shamoto, T.C. Hsu, W.K. Hong (1982) Reduction of adriamycin-toxicity by ascorbate in mice and guinea pigs. *Cancer Res* 42: 309–316.

12. Garewal, H. (1995) Beta-carotene and antioxidant nutrients in oral cancer prevention. In Prasad K.N., Santamaria L., Williams R.M. (eds) *Nutrients in Cancer Prevention and Treatment* Totawa, N.J.: Human Press, pp. 235–247.

13. Gey, K.F. (1993) Prospects for the prevention of free radical disease, regarding cancer and cardiovascular disease. In: Free Radicals in Medicine. *British Medical Bulletin* 49(3): 679–699.

14. Griffiths, K. (1996) *Nutrition and Cancer.* Oxford, U.K. Isis Medical Media: Lakefield, Aylesford, Kent, U.K.

15. Gundimeda, U., S. Hara, W.B. Anderson and R. Gopalakrishnakone (1993) Retinoids inhibit the oxidation modification of protein kinase C. induced by oxidant tumor promoters. *Arch Biochem. Biophys.* 300: 576–630.

16. Kim, J.H., S.H. Kim, S.Q. He, J. Dvagovic and S. Brown (1995) Vitamins as adjuvant to conventional cancer therapy. In Prasad K.N., Santamaria L., William R.M. (eds): *Nutrients in Cancer Prevention and Treatment. Totawa, N.J. and Humana Press,* pp 363–372.

17. Kline, K., W. Yu and B. Zhoa (1995) Vitamin E succinate: Mechanisms of action as tumor cell growth inhibitor. In Prasad K.N., Santamaria I., William R.M. (eds). *Nutrients in Cancer Prevention and Treatment,* Totawa, N.J: Humana Press, pp 39–55.

18. Krauhausen, U., U. Blum and H.P. Fortmeyer (1989) Vitamin B6 responsive growth of human lung cancer in nude mice. *Strhlentherapie und Onkologie* 1665: 562–563.

19. Labriola, D. and R. Livingston (1999) Possible interactions between dietary antioxidants and chemotherapy. *Oncology* 13: 1003–1012.

20. Lamson, D.W. and M.S. Brignall (1999) Antioxidants in cancer therapy: Their

actions andinteractions with oncologic therapies. *Alternative Medicine Reviews* 4(5): 304–329.

21. Mahoney, C.W. and A. Azzi (1998) Vitamin E. inhibits kinase C activity. *Biochem. Biophys. Res. Comm.* 154: 694–697.

22. Park, C.H. (1998) Vitamin C in leukemia and preleukaemia cell growth. *Prog. Clin. Biol. Res.* 259: 321–330.

23. Prasad, K.N. (1980) Modulation of the effect of tumor therapeutic agents by vitamin C. *Life Sciences* 27: 275–280.

24. Prasad, K.N. and J. Edwards-Prasad (1990) Expression of some molecular cancer risk factors and their modification by vitamins. *J. Am. Coll. Nutr.* 9: 28–34.

25. Prasad, K.N., J. Edwards-Prasad, S. Kumar and A. Meyers (1993) Vitamins regulate gene expression and induce differentiation and growth inhibition in cancer cells. Their relevance in cancer prevention. *Arch Otolaryngol Head Neck Surg* 1189: 1133–1140.

26. Prasad, K.N., C. Hernandez, J. Edwards-Prasad, J. Nelson, T. Borus and W.A. Robinson (1994) Modification of the effect of tamoxifen, cis-platin, DTIC and interferon-α2b on human melanoma cells in culture by a mixture of vitamins. *Nutr Cancer* 22: 223–245.

27. Prasad, K.N. and R. Kumar (1996) Effect of individual antioxidant vitamins alone and in combination on growth and differentiation of human non-tumorigenic parotid acinar cells in culture. *Nutr. Cancer* 26: 11–19.

28. Prasad, K.N., A. Kumar, V. Kochupillai and W. Cole (1999) High doses of multiple antioxidant vitamins: Essential ingredients in improving the efficacy of standard cancer therapy. *J. Amer. Coll. Nutr.* 18(1): 13–25.

29. Seifter E., A. Rettura, J. Padawar and S.M. Levenson (1984) Vitamin A and beta-carotene in adjunctive therapy to tumor excision, radiation therapy and chemotherapy. *In Prasad K.N. (ed): Vitamins, Nutrition and Cancer. Basel: Karger,* pp 1–19.

30. Sonnevald, P. (1978) Effect of alpha tocopherol on cardiotoxicity of adriamycin in the rat. *Cancer Treat. Rep.* 62: 1033–1036.

31. Trizna, Z., S.P. Schantz, J.J. Lee, M.R. Spitz, H. Goepfert, *et al.* (1993) *In vitro* protective effect of chemopreventive agents against bleomycin-induced genotoxicity in lymphoblastoid cell lines and peripheral lymphocytes of head and neck cancer patients. *Cancer. Detect. Prev.* 17: 575–583.

32. Turley, J.M., F.W. Ruscetti, S. Kim and T. Fu (1997) Vitamin E succinate inhibits proliferation of BT-20 human breast cancer cells: increased binding of cyclic AMP negatively regulates E_2F transactivation activity. *Cancer Res.* 57: 2668–2675.

33. VandeCreek, L., E. Rogers and J. Lester (1999) Use of alternative therapies among breast cancer outpatients compared with the general population. *Alternative Therapies* 5(1): 71–76.

34. Yamanaka, N., M. Fukishima, K. Koizumi, K. Nishida, T. Kato and K. Ota, (1980) Enhancement of DNA chain breakage by bleomycin and biological free radical producing systems. *In DeDuve C and Hayashi T (eds): Tocopherol, Oxygen and Biomembranes. New York: N Holland,* pp. 59–69.

35. Zhang, I.X., R.V. Cooney and J.S. Bertram (1992) Carotenoids up-regulate connexin-43 gene expression independent of their provitamin A or antioxidant properties. *Cancer Res.* 2: 5705–5712.

Radiobiology and Bio-Medical Research
Edited by K.P. Mishra
Copyright © 2004 Narosa Publishing House, New Delhi, India

5. New Vistas in Nuclear Medicine

A.M. Samuel

Biomedical Group, BARC, Mumbai 400 085, India

Giant strides in technology and its application in medicine in the past three to four decades has left one breathless, stunned, as well as excited. The older concepts of medicine wherein diagnosis was dependant on a careful history taking, examination by a skilled physician, biochemical investigations which at best were resultant of abnormal metabolic processes, or imaging of body organs by X-rays with or without contrast media along with a keen intellect were used in arriving at a probable diagnosis of the disease and thus leading to institution of appropriate therapies.

Rapid advances in imaging techniques have occurred in the last few years. The process of obtaining images of body organs have used a varied range of principles of physics, such as ultrasound, proton resonance and gamma radiation. This has revolutionized the ability and accuracy of diagnosis of diseases. New diseases have been detected and older ones have been understood in a more detailed manner. Imaging with improved versions of machines like computerized tomography, magnetic resonance imaging and ultrasound have enabled one to visualize lesions in organs with resolutions as small as 2 mm such that early diagnosis can be made and treatment can be instituted resulting in better results for the patients. There is a constant evolution of the machines with the intention to improve resolution, image processing with software, and data analysis so as to enhance the diagnostic skills of the physician. The sophistication of the imaging devices for better resolution is of course translated in increased costs. A stage may be reached where the cost benefit ratio may demand consideration of other options and facilities.

The evolution towards this goal has begun. The need will be to not only detect very small lesions on images but to develop methods of measuring biochemical changes in the organ resulting from disease processes. This is presently obtained with MRI spectroscopic machines wherein phosphorous in molecules like ATP, ADP and others can be determined. Tumors have shown characteristic differences from normal tissues. Hence added information about tissue level changes along with improved imaging abilities of the machines adds to diagnostic efficacy. It is anticipated that this trend will progress further. These technologies have enabled one to study tissues and the changes which occur therein. Its like looking down a microscope in a living being.

Is it possible to look still further into the cell, the many components of the cell and to study metabolic processes taking place within the cell? The answer is

YES. We are now in an era described as molecular medicine. This is feasible due to the availability of molecules like F18 deoxyglucose, a radiolabelled metabolic substrate which mimics glucose in the body. With the help of this molecule one can study metabolic functions of the tissues and cells. C11 labeled amino acids, lipids and proteins enable one to assess the metabolism of proteins and fats. Receptors on cell membranes and intracellular ones can be studied by using appropriately labeled compounds. So it can be seen that the receptors of cells, the metabolism of nutrients of cells can be, as it were, studied in real time in the living body. A prospect which could not have been imagined a decade ago.

Nuclear medicine, a unique discipline in medicine, over the last four decades has evolved so rapidly that it has become difficult to keep abreast of the advances that are taking place at an exponential growth rate. Despite its great potential and useful applications in the diagnosis and management of patients it is still a relatively unknown entity.

The unique contribution of this specialty is the ability to examine the dynamic state of body constituents as reflected in every organ of the body. The emissions of gamma photons, which can be measured by radiation detectors outside of the human body has enabled the study of regional function and its underlying biochemistry. In fact the advances in this field are such that it may be necessary to change the term 'Nuclear Medicine' to a more appropriate term which will indicate the true potential of the discipline to 'Molecular Medicine'. In future, measurement of *in vivo* chemistry will become commonplace. Molecules will be the targets of detection instead of cells, tissues and organs, in other words we will start looking inside the cells rather than at the cells.

An excellent article by Dr. Henry Wagner (designated as the father of nuclear medicine [1]) has described the future of nuclear medicine as a very promising and exciting one. He has defined disease as a disorder of chemistry of the body. Nuclear medicine techniques are the most sensitive and appropriate methods to characterize the biochemical changes that occur as a result of disease processes. More than ever before will nuclear medicine physicians have to answer questions as to how specific nuclear medicine procedures help in solving clinical problems by determining the changes that occur in the cell and result in dysfunction of the organ, to assess the right strategy for treatment, and to help in designing drugs which will be most effective at the cellular level.

The use of radioactive tracers in medicine is comparable to the invention of the chemical balance and the discovery of X-rays. The tracer principle, that is, the monitoring of molecules that participate in the dynamic state of body constituents, led to a whole new approach to biology and medicine, characterizing the chemistry of growth, development, and maintenance of life in experimental animals and human beings. No other techniques have the sensitivity, specificity, and quantifiability in the study of in situ chemistry. No other field of medicine has a greater ability to define disease as a problem in coordinating and balancing the billions of chemical reactions involved in the normal function of cells and tissues in the body. Historically, whenever a biological or clinical finding is made, whether the finding is physiological, pathological, or clinical, scientists interpret the observed phenomenon in terms of chemical reactions. As modern medicine becomes oriented towards molecules rather than cellular, nuclear medicine will be in the forefront and will become an integral part of medicine.

Advances in nuclear medicine will proceed along two principle lines, the development of improved sensitive detectors of radiation, powerful and interpretable data processing, image processing, display systems; and the production of exotic and new but useful radiopharmaceuticals.

One of the most interesting and illustrative examples of the progress in this direction is that of the study of dopamine receptors in Parkinson's disease. Almost 15 years ago, in1983, one of the first scans performed with radiolabeled compound, which could bind to dopaminergic receptors was performed by Dr. Henry Wagner on himself. The demonstration of these receptors in the basal ganglia and later the ability to quantify the concentration of the receptors has played a significant role in understanding the pathophysiology of the disease and to assess the effect of different drugs in the management of the disease. The progress is such that it is now possible to examine all aspects of dopamine neurotransmission. The synthesis of dopamine with F18 levodopa, the secretion of dopamine from the presynaptic neurones, the metabolism of dopamine by the monooxidase enzyme system, and the reuptake of unbound dopamine from the synapse by the presynaptic dopamine transporter. Such measurements can help in differentiating various movement disorders like senile Parkinson's disease, striatonigral degeneration, traumatic Parkinsons, progressive supranuclear palsy, and idiopathic Parkinsons disease. In classical Parkinsons disease there is a deficiency of the dopaminergic presynaptic neurones, while in idiopathic Parkinson's disease the D1 and D2 receptors are normal and in striatonigral there is degeneration of postsynaptic neurones. This indicates that the drug therapy is to be tailored to the biochemical changes occurring in the diseases. Since the clinical manifestations of Parkinson's is related to the degree of the receptor deficiency, monooxidase inhibitors can be useful in the early stages of the disease. The availability of positron and single photon emitting radiotracers to study dopamine transporters can be helpful in the study of environmental and genetic factors associated with movement disorders. This is only an example of how nuclear medicine or molecular medicine has changed the world of physicians in understanding the very basis of diseases which were considered as black boxes and treatment was based on a hit and miss strategy. This regional molecular approach to diagnosis can be applied in several neuropsychiatric disorders, cognitive disorders, depression, and many others.

Another important disease to be understood at the molecular level is cancer. The definition is wide ranging depending on which aspect of cancer one looks at. Whether it is the unruly and autonomous growth of the cells or the function and many other cellular and molecular aspects of the disease. However, one such definition is of the view that cancer is a communication disorder. Life is maintained because atoms and molecules can recognize each other. Radioactive tracers can be 'molecules with messages' such as mRNA, then as hormones and neurotransmitters and message signals, information transfers, recognition of molecules by receptors, transporters of molecules, all of which has evolved from unicellular organisms as a means of facilitating intercellular molecular communication.

Many types of cancers can be detected by the expression of receptors on plasma membranes such as VIP (vasoactive intestinal peptide) or somatostatin {SS}, and others. Examples of molecular recognition units, the biological active peptides that modulate cell proliferation and the vast potential for the possible positron or single photon labels that can be used to study the molecules in different cancers are increasing day by day. Metabolic changes in cancers occur usually by virtue of up regulation of glycolytic pathway and increased expression of glucose transporter proteins allowing the study of this pathway by F18 flurodeoxyglucose and PET imaging. Intensive studies on the GLUT 1-6 transporters and the inhibitors is under way so that appropriate compounds can be synthesized and used to detect the receptors on the plasma membrane of various cancers. Growth factors and cytokines like EGF (epidermal growth factor) and IGF (insulin like growth factor) have important proliferative effects in various cancers and can be blocked by monoclonal antibodies which can react specifically to the selected epitopes on the receptors. Hence many monoclonal antibodies have been developed and are under investigation to detect and treat cancers which express these growth factors. For better detection of neoplasms by scintigraphy, ligands that bind specifically to surface receptors overexpressed in tumor cells are the potential candidates. The promising ligands include bioactive endogenous peptides or their analogues, inhibitors of glucose transport proteins, estrogen and sigma receptor ligands, growth factors and cytokines.

At the dawn of the 21st century, nuclear oncology is undergoing a formidable and rapid mutagenesis. The progress in radiochemistry, radiopharmacy and foremost, the advances in molecular oncology are the determinant factors. The astonishing developments in peptide and nucleic acid chemistry have enabled the development of newer, highly specific probes such as antisense, aptamer and peptidogenic molecules so as to image the oncogenes and antioncogenes, transcriptional (messenger ribonucleic acid) and translational (protein products) involved in carcinogenesis. In fact the concept of imaging genes with nuclear probes has become a virtual reality.

Indeed we are witnessing an interesting age of technology which allows one to peer into one's body, into the cell, how it functions and at the DNA itself. It almost appears like the science fiction movies one sees today. What is fiction today will translate to a reality tomorrow.

Aside from the new radiopharmaceuticals which are likely to be developed there will be better and improved models of detectors. The four recent advances in PET instrumentation are in gantry innovations and in new detector designs. The future prospects for detector designs is to have a system that has individual coupling of the scintillator to a solid state readout system and to phase out the photoelectron multiplier. An extensive search is underway to find a scintillator with improved light output, speed and efficiency. Cerium doped lutetium orthosilicate has been shown to have most of these properties and new machines may be made of this scintillator [3]. Advances in three dimensional reconstruction has led to the development of a commercial, rotating, partial ring, fully 3-D positron emission tomography scanner [2].

Another scope for future developments is in therapy or treatment of cancers and non-cancerous disorders with appropriate radionucleides. The use of Monoclonal antibodies labeled with alpha or beta emitters is being tested extensively. The initial euphoria of discovering the magic bullet has not stood up to its promise. There are many problems and each of which has to have a solution. This has taken its toll in that the treatment with monoclonal antibodies is not being used as extensively is it should have been. However, these compounds still hold promise and persistent efforts to overcome the hurdles is underway. Peptides, receptors, and other ligands are also under trials. Many new radiopharmaceuticals tagged with a variety of exotic radionucleides like europium, holmium and others are being tested for pain relief. Recent reports suggest that intraluminal irradiation of coronary arteries in conjunction with balloon angioplasty reduces proliferation of smooth muscle cells and neointima formation, thereby inhibiting restenosis. A variety of radioisotopes are being tried among them are Y90 chloride, P32 and others. The studies are in the preliminary stage but appear promising. Treatment of coronary artery disease and other vascular disorders which were not amenable to surgery or stenosed again have an excellent chance of being treated. So it seems that nuclear medicine techniques not only look inside the cell at metabolic activities but inside the small blood vessels and help in maintain the patency of the vessels. All this almost appears like a science fiction novel. The unbelievable advances in science have occurred in a short span of our own lifetime.

Ten predictions for the future described by Henry Wagner [4] in brief, are as follows:

- Just as molecular genetics will revolutionize the practice of medicine, so also will nuclear medicine revolutionize the practice of molecular genetics.
- Measurement of *in vivo* chemistry will become as prevalent as biochemical examination of blood and urine is today.
- Abnormal regional chemistry will affect the design of drugs and the planning and monitoring of drug treatment.
- Chemotyping will link genotyping and phenotyping.
- Homeostatic processes characterized by radiotracers will make it possible to to examine the genetics of homeostasis, repair processes and other functions.
- Gene mutations associated with diseases and the way the abnormal gene functions.
- Study of breakdown of feedback processes and the resultant disease manifestations can be evaluated.
- Targeting of specific drugs to specific cancer causing genes and cancer causing mutations can be studied with tracers.
- Nuclear medicine will change the concepts of disease and will provide a new way of looking at diseases.
- Detection of chemical correlates of behavioral disorders, as well as somatic diseases and will focus on neurobiological, socio-economic and political factors in neuropsychiatric disorders.

An apt quotation from Lincoln "If we can determine where we are and wither we are tending, then we will best know what to do and how to do it."

A Look Into the Crystal Ball for the Future

- Integration–networking of disciplines
- Genetic engineering
- Neurosciences
- Microelectromechanical machines
- Combinatorial chemistry
- Nanoscale engineering
- Transgenics
- Chip technologies for diagnostics

The coming century can be called as the 'CENTURY OF THE BRAIN'
  Feynmans prediction —'QUANTUM COMPUTERS'

References

1. Amols, H.I., L.E. Reinstein and J. Weinberger (1996) Dosimetry of a radioactive coronary balloon dilation catheter for treatment of neointimal hyperplasia. *Med. Phys.* 23: 1783–1788.
2. Bailey, D.L., H. Young and P.M. Bloomfield, *et al.* (1997) ECAT ART–a continuously rotating PET camera: performance characteristics, initial clinical studies, and installation considerations in a nuclear medicine department. *Eur. J. Nucl. Med.* 24: 6–15.
3. Budinger, T.F., K.M. Brennan, W.W. Moses and S.E. Derenzo (1996) Advances in Positron Tomography for Oncology. *Nucl. Med. Biolog.* 23: 659–667.
4. Wagner, H.N. (1996) The future. *Semin. Nucl. Med.* 26: 194–200.
5. McAfee, J.G. and R.D. Neumann (1996) Radiolabeled peptides and other ligands for receptors overexpressed in tumor cells for imaging neoplasms. *Nucl. Med. Biolog.* 23: 673–676.

Radiobiology and Bio-Medical Research
Edited by K.P. Mishra
Copyright © 2004 Narosa Publishing House, New Delhi, India

6. Emerging Scenario in Radiation Biology of Mammalian Cells: Prospects and Opportunities in the New Millennium

K.P. Mishra

Cellular and Free Radical Radiation Biology Section, Radiation Biology Division,
Bhabha Atomic Research Centre, Mumbai 400 085, India
kpm@magnum.barc.ernet.in

Introduction

Humankind has always been exposed to natural radiation since radioactivity and emanating ionizing radiation existed on the earth long before life emerged. The discoveries of X-rays by Wilhelm Wrontgen and radioactivity by Henri Bequerel at the end of 19th century have opened up an exciting new world of science. The quest to unravel the innermost secret of matter revealed enormous amount of energy stored in an atom. This new source of atomic power was soon applied to generate electricity, to develop nuclear weapon and to treat cancer. Over the past decades nuclear and radiation technologies have proved valuable tool for addressing a range of needs and problems of people. Several important achievements have been made using nuclear methods in the field of energy, medicine, industry and human health. Exponential growth has occurred in biomedical research and applications by using radioisotopes for diagnosis, imaging and cancer treatment. Early investigators, namely, Becquerrel, Marie Curie and others recognized the harmful effects of radiation on human body and the field of radiation biology was since born. In her thesis on physiological effects of radium, Marie Curie described observations on skin, damaging effects on plants, inhibition of growth of microbes and killing of mice [6]. Modern medicine has been revolutionized with the use of radiation or radioisotope for diagnostic radiology and radiotherapy. In addition, over the years experiments have shown that high doses of atomic radiations either natural or accidental produced harmful effects to an exposed individual. Effects of radiation became a subject of urgent importance after explosion of atomic bomb at Hiroshima and Nagasaki in Japan at the end of Second World War in 1945. The horror of destructive power of atomic bomb shocked the world but prompted new impetus for studies on health consequences of radiation exposure. Ionizing radiation and nuclear energy have found diverse peaceful applications and immense interest has been generated in professionals as well as among general public to understand biological effects of ionizing radiation related to health effects and risk to people.

Ionizing radiation can potentially induce cancer and genetic effects which has been subject of active investigation in the past years. In view of the importance of radiation effects on humans, the UN General Assembly has set up United Nations Scientific Committee on the Effects of Atomic Radiation (UNSCEAR) in 1955, which periodically produces elaborate reports on doses, effects and risks of radiation on a world-wide scale from both human-made and natural radiation sources. One of the main forums devoted to task of health protection from ionizing radiation is the International Commission on Radiation Protection (ICRP) founded as early as in 1928. In the rapidly growing scientific literature a wealth of factual information, new ideas and working models have been accumulated on biological effects of ionizing radiation including detrimental effects on human health. Many International and scientific organizations are involved in the development of radiation protection standards, regulation of safe use of nuclear radiation and practical implementation of rules and recommendations.

Ionizing radiation and radioisotopes have become a familiar speciality proving a powerful tool in hospitals to diagnose ailments and treat diseases including cancer. In addition, nuclear energy has found applications in diverse areas ranging from medical research to weapon development. It has become an imperative social responsibility to understand the basic mechanisms by which ionizing radiations interact with living matter to produce the observed effects. Radiation biology aims at understanding the biological effects of ionizing radiation with a hope to provide basis for precise evaluation of biological risks involved and optimization of gainful uses of radiation in research, medicine, industry, military and agriculture. The use of ionizing radiation in study of cancer and its treatment owes heavily to contributions of radiobiology. Full understanding of basic concepts and mechanisms of radiobiological practice for formulation of radiation protection strategies has become an urgent requirement for safety of individuals, progeny and human race. Radiation biological research has made important contribution to determine dose level for population, occupational dose effects and maximum permissible dose limits.

In this talk I wish to present a brief account of basic radiobiological principles with relevance to application of radiation and radioisotopes for improvement of quality of life of common people. I also intend to outline the goals and challenges of radiobiology in the new millennium with an emphasis to develop new research strategies to tackle the problems faced in developing radiation related technologies.

Chemical Basis of Radiobiology

The physical absorption of radiation energy in living biological cells is followed by initiation of a complex free radical reaction consequential to the life of a cell. Long before the molecular nature of radiation damage to biological material became known, it was generally assumed that the damage reflected a combined effect of the direct action of radiation on critical cellular molecules and indirect effects resulting from chemical reactions of vital molecules with free radicals generated in surrounding water. Subsequent radiation chemical investigations using biophysical techniques have largely confirmed involvement of radiation generated free radicals in biological damage [26, 19]. Extensive studies showed a chemical basis for understanding the mechanism of radiosensitizing effects of oxygen. This stimulated further research and a variety of compounds and drugs such as N-ethylmaleimide and iodoacetamide etc have been shown to modify radiation damage to living tissues by interfering in radiation chemical reactions [26, 38]. Numerous studies examined radiochemical effects of molecular oxygen elucidating the mechanism of oxygen mediated fixation of damage [40, 7].

Damaging effects of ionizing radiation are produced by transferring its energy directly to the target molecules of cells or by radiolysis of water present in the surrounding. Depending on the energy of a particular radiation, they deposit their energy in the medium by photoelectric effect, compton scattering and pair production processes. The depth of tissue penetration of particulate radiation depends on charge, mass and initial energy. Both electromagnetic and particulate radiations produce random excitations and ionizations in the atoms or molecules of the absorbing medium. Each radiation absorptive event in the living tissues has potential to culminate in a biological consequence by breakage of chemical bonds generating radicals which initiate reactions at the biochemical level. The indirect action of radiation consists in generation and reactions of primary water radicals, namely, hydrated electron (e_{aq}^-), hydrogen atom ($\cdot$OH), hydroxyl radical ($\cdot$OH) which contribute in causing damage to biological systems [4]. A dose of gamma radiation produces many chemical lesions in cellular molecules such as nucleic acids, proteins, lipids and carbohydrates and approximately two thirds of it is accounted for by indirect effect mainly by hydroxyl radicals. Only a small fraction of this chemical damage leads to injurious biological effects. The extent of $\cdot$OH mediated damage is determined by the product of its rate constant and relative mole fractions of the cellular components. The radiation generated water free radicals react either with a biomolecule or with each other giving molecular products, (H_2, H_2O_2, H_2O) or may be removed by reaction with natural scavengers. The complex reactions of water radicals with structural cellular molecules may culminate into malfunctioning of cells. The free radical initiated cell injury is known to be remarkably enhanced by the presence of oxygen. This ability of oxygen to enhance the effectiveness of a dose of radiation is believed to be due to fixation of damage by efficient reaction of oxygen with radiation induced primary or secondary radicals in the system. These physico-chemical events occur within a millionth of a milli second to minutes, hours and days after exposure to ionizing radiation.

Extensive studies on radiation chemical and biological effects have provided basis to understand the nature of lethal radiation lesions in mammalian cells. These studies have provided a better molecular insight into processes such as fixation, repair and modulation of radiation induced damage by endogenous as well as exogenous modifiers. Reactions of these radicals in aqueous system have been extensively investigated by spectroscopic techniques, which provided a basis for interpretation of radiation chemical mechanisms of cellular response to ionizing radiation [25, 15]. Present state of knowledge is, however, far from complete and there exists a wide gap in our understanding of the link between physicochemical changes and biological effects after irradiation.

Cellular Radiobiological Effects: Current Understanding

Background
In the middle of this century studies on response of normal and tumor cells entered a new phase with development of assay for measuring the clonogenicity of individual cells. Significant insight into the effects of radiation on mammalian cells have been obtained by using cell-culture systems. These assays enabled researchers to move from measuring gross tissue response to survival of cells in critical cell population. Puck and Marcus published their results on changes in HeLa cell viability as a function of radiation dose [35]. Survival curve for tumor cells irradiated *in vivo* was obtained by Hewitt and Wilson (1959) using a transplanted lymphocytic leukemia which produced lymphocyte infiltrate ions in the liver of mice. The similarity of resultant dose response curve to the survival curve in HeLa cells in culture provided evidence that *in vitro* findings obtained using established cell lines were relevant to understanding the *in vivo* effects of radiation. The very shape of survival curve yielded many new information. Experimental survival curve fitted well with the mathematical theory which rested on the assumption that cell contained certain critical structures (targets) and it would die when all of these structures were destroyed by radiation which is based on the concept of random inactivation of identical targets by radiation. The shoulder on the curve was interpreted in terms of multiple targets and that cells were capable to accumulate and tolerate sublethal radiation damage without being killed. Implications for these results were enormous for translation in cancer radiotherapy. The fact that cells were able to accumulate and tolerate a certain amount of sublethal damage provided a theoretical basis understand to the protection seen with fractionation and low dose rate irradiations. Different cell lines showed differences in their radiosensitivity reflecting their ability to repair the radiation damage, which was verifiable in different types of normal and malignant cells *in vivo*. These results provided a rationale for fractionation regimens in cancer radiotherapy. They also stimulated interest in studies on high LET radiation effects on cells which were predicted to produce less repairable damage with far reaching implications in tackling the problem of commonly observed radio-resistance in cells.

Modes of cell death
Ionizing radiation interaction with living cells is central to modern radiation biology. In recent years many new information, models and ideas have been accumulated concerning biological effects of ionizing radiation. Mechanism of cell death has acquired considerable significance in the past some years. Irradiated cells die either by necrosis or by a genetically programmed mode of cell death called apoptosis. Kerr *et al.*, (1972) characterized important differences between these two modes of death. Necrosis is a passive form of cell death which involves metabolic collapse following massive damage to cell especially to membrane allowing swelling, leakage of intracellular content to surrounding and producing inflammatory response. In contrast, apoptosis is an active mode of cell death— genetically programmed self-destruction of the cell (physiological cell death)

which is manifested by shrinkage, membrane blebbing, chromatin condensation and nuclear fragmentation. Apoptosis involves well-regulated mode of cellular response to external or internal stimuli. Cells dying *in vivo* maintain their membrane integrity until they have been cleared by phagocytosis. However, *in vitro* cells lacking phagocytes ultimately result in plasma membrane breakdown, a phenomenon called *secondary necrosis*. Radiation biologists describe cell death in terms of reproductive and interphase modes; former is characterized by loss of cell's ability to undergo unlimited cell division (proliferation) forming a colony and cell death occurs at or after first mitosis following irradiation while the later is described by impairment of cellular metabolism and disintegration of cells before entering mitosis. Relationship between necrotic and apoptotic process is far from simple. Currently much research has been devoted to understand basic mechanism of apoptosis in normal and tumor cells because of accumulated knowledge that many factors that control initiation and progression of tumor also modulate apoptotic potential of cells. Radiobiological research is at the door of making significant advancement in deeper understanding of basis of cacinogenesis and tumorigenesis.

Dose-response relationship
Associated with the molecular mechanism of cell killing is a most hotly debated question at scientific forums: dose-response relationship. What has remained unresolved is whether it is justified to extrapolate linearly the effects from large doses to low doses including doses in the range of natural level of radiation. Essential features of dose response relationship in single cell irradiation fits linear quadratic mathematical relationship designated as deterministic and stochastic. Deterministic effects occur above certain threshold in moderately high dose range (above 500-1000 mSv). Severity of the damage depends on the absorbed dose. The effects are detectable by laboratory or clinical methods. On the other hand, stochastic effects occur at low doses of radiation (100-200 mSv). Probability of the consequence increases with dose and relationship between dose and effect is assumed to be linear. The model for assessing detrimental health effects non-linear threshold model is based on deterministic effect (NL-T) while for stochastic effects linear-non-threshold model is used. Not having a threshold, stochastic model stipulates that, however, small but certain amount of risk exists to human health even at the lowest dose of exposure. The dilemma is whether health risks exist at low doses and future research work will have to provide an answer.

Cellular targets
A great deal of research on the mechanism of cell killing by ionizing radiation considers genomic DNA as central target and repairability of DNA damage especially double strand break determines restorability of cell to preirradiation state. However, advances in molecular radiobiology in recent years revealed that in addition to causing DNA damage, radiation affects membrane and organelles producing alterations in signaling, gene expression, cellular redox reactions and cell-cycle regulation. In multicellular organisms, radiation mediated damage

occurs at different level of biological organization and the major radiation effects seem to be mediated by indirect actions of free radical generation in water, which subsequently react with vital biological molecules producing changes such as genetic effects, cell killing and carcinogenesis. Chemically reactive free radicals generated by radiation in cellular molecules or water may damage cell function or prevent it from reproducing altering genetic information or potentially kill the cell. Biological effects on cells, therefore, require investigations at various level of cellular organization. To explain the radiation inactivation of molecules and cells, the concept of 'target theory' was developed in 1940s, which supposes certain critical molecules or 'targets' present inside the cells that must be hit/ inactivated by radiation to kill the cell. To some extent this theory explains the radiation damage to mammalian cells, unicellular organisms and inactivation of molecules like protein in solutions. However, this approach seems inadequate for complex living cells which display secondary biological processes such as repair and recovery. The radiation damage to cellular DNA generally produces single strand breaks (SSB) at low dose range (~1 Gy) and at higher doses (10-100 Gy) double stand break (DSB) are frequently observed. Radiation mediated DNA damage is believed to linked to delayed DNA synthesis and inhibition of cell division. Undoubtedly, DNA is a critical radiation target but experimental data point to a gap between DNA repair and cellular radiosensitivity. Membrane radiation damage may become significant at non-lethal or moderate and therapeutic doses of radiation. More sensitive and new may help to determine whether or not membrane is another primary target for radiation cell killing.

Experimental examination of two end points, namely, DNA damage and cell survival present intriguing gap and understanding of mechanism of cell killing by radiation is far from complete. DNA repair has been considered to play a role in radiosensitization of cells but recent findings contradict the most prominent model of cellular radiosensitivity and favor an explanation in terms of α/β model of cell survival as a function of radiation dose.

Membrane Radiosensitivity and new Opportunities

Cell membrane as radiation target
Right from the beginning understanding the effects of ionizing radiation on cell membrane has been considered important especially changes in their permeability properties, for example, enzyme release theory. Over the years, progress in deeper insight of membrane structure and function, membrane radiobiology has received a new perspective and the membrane target theory has come into renewed focus. Literature is growing with convincing evidence that events associated with membrane damage play a role in the development of radiation injury in cells. In past years, vulnerability of membrane as another important radiation target in cell was suggested providing plausible explanation of radiation 'oxygen effect' in the development of oxidative damage of membrane [1]. Based on this work on radiation effects on microorganisms, a hypothesis was put forward advocating existence of at least two main types of lesions in cells, namely, 'N' type (mainly associated with DNA) and 'O' type damage (mainly argued in favor

of membrane damage). These suggestions were made in a background of limited information on structure and component of cellular membranes and little was known about involvement of membrane in molecular signaling and apoptotic death of cells. A generally considered argument in favor of DNA as primary and exclusive target for radiation found increasing support because of its central role in transmission of genetic information. However, complex nature of membrane enclosing vital cellular constituents legitimately point to vulnerability and uniqueness in radiation damage to cells. Contributions of membrane damage in determining the cellular response to radiation exposure need urgently a fresh evaluation. With the availability of sensitive biophysical and molecular techniques, alteration in function and properties of membrane has become possible to be detected after exposure to relatively low doses of radiation. However, the correlation of these changes with vital cellular processes such as lethality of cells remains to be firmly established.

Free Radical and Lipid Peroxidation
Recent results on damaging events associated with membrane such as peroxidation, and molecular transport strongly suggest a role of membrane in several cellular processes including induction of cancer. The hypothesis of membrane target finds favor from the fact that many factors that alter membrane function also modify cellular response to radiation exposure. A number of articles have been published strongly advocating the role of cellular membrane as a sensitive and critical target in the mechanisms of radiation damage to tissues/cells [1, 14, 37, 7, 32]. Two major constituents of biological membranes, namely, lipids and proteins are prone to oxidative damage due to presence of unsaturated bonds in lipids and perhaps the lipid (also protein) peroxidation constitutes a major event in radiation induced membrane damage. Alterations in surface charges, membrane bound enzymatic activity and receptor function after cellular radiation exposure has been documented in literature [14]. Extensive investigations have shown the involvement of molecular damage to membrane in the loss of cellular function after radiation exposure [15, 18, 20, 7]. Mechanism by which free radicals generated by radiation cause oxidative damage to membrane lipids and modify properties of cellular membrane has been receiving increasing attention [16, 33]. Radicals of lipid (L) are believed to be formed by its reaction with water free radicals ($\cdot$OH) produced by ionizing radiation. Polyunsaturated fatty acids (LH) represent a sensitive molecular site which react with oxygen to form lipid peroxyl radical (LOO) after undergoing molecular rearrangements of conjugation in double bonds and eventually a chain reaction sets in following irradiation of oxygenated samples [40]. Studies on radiation effects on cellular membrane have demonstrated generation of free radicals in lipid and lipoprotein molecules of membrane leading to peroxidative chain reactions, alteration in lipid bilayer fluidity and modification of membrane protein channel properties. The lipid peroxidation induced alterations in cellular functions have been reported involving inactivation of membrane enzymes, inhibition of cell division and cellular proliferation. Moreover, changes in membrane fluidity at high and low doses of radiation reflect radiation oxidative damage, which may serve as an indicator of radiation injury.

Increased production of reactive oxygen species (ROS) is a feature of many human diseases including cardiovascular diseases and cancer. Cells can respond to mild oxidative stress by up-regulating antioxidant defenses and other regulatory systems, but severe stress can damage DNA, proteins and lipids leading to cell transformation or death by apoptosis or necrotic mechanism [11]. Many of the chemical and physical stimuli capable of inducing apoptosis are known to evoke oxidative stress by increasing steady state concentration of ROS. The formation of ROS in a stressed cell occurs as a consequence of mitochondrial consumption of oxygen, hydrogen peroxide production in peroxysomes, respiratory bursts during activation of phagocytes and during induction of cytochrome P450 enzymes. The ROS appear non-specific but they can induce and mediate coordinated events in apoptosis.

Oxidative Membrane Damage and Apoptosis

Process of apoptosis in cells is proving to be of great interest in cancer biology, which seems linked with damage to membrane. Radiation injury to both plasma and nuclear membrane initiate pathways for induction of apoptosis. It has been demonstrated that gamma rays induce apoptotic changes in bovine epithelial cells without indication of direct damage to their DNA [10]. Involvement of membrane related signaling mechanism has been demonstrated through radiation induced hydrolysis of sphingomyelin generating ceramide in plasma membrane. It has been shown that inhibition of ceramide formation significantly prevented apoptosis. More recent results have demonstrated that lipid peroxidizing agents such as hydrogen peroxide, t-butyl hydroperoxides and fatty acid hydroperoxides could trigger apoptotic cell death in various cell system [5]. Protein kinase C activation was found to block both radiation induce sphingomyelin hydrolysis and apoptosis. The role of plasma membrane in irradiation induced apoptosis was further demonstrated by studies on thymocytes [36]. Pre-irradiation treatment of mouse thymocytes by Trolox (a vitamin E analogue protecting membrane damage) could not only protect the cells from apoptosis but also blocked the influx of calcium, which is known for activation of endonucleases involved in apoptotic cell death. The gamma irradiation of liposomal model membrane has shown significantly enhanced membrane rigidity possibly due to cross-reaction of lipid radicals. Results have shown that modifications of physical properties of membrane by inclusion of cholesterol in liposomal composition produced inhibition of membrane oxidative damage, which has been ascribed to slow chain propagation reaction in lipid peroxidation reaction [32]. Free radical mediated damage to plasma membrane of mouse thymocytes that its inhibition in presence of antioxidants inhibit the membrane damage suggesting involvement of free radical mechanism in the damage to membrane [33, 34].

The ability of ionizing radiation to induce apoptosis, and the associated changes in membrane properties and nuclear structure has been shown in thymocytes, intestinal crypt cells and lymphocytes. Permeability of plasma membrane of mouse thymocytes have been shown to significantly increase following gamma irradiation at moderate doses of radiation (10 Gy), as measured by fluorescence method (Pandey and Mishra 2000 unpublished result). It also been demonstrated that these membrane changes were correlated with nuclear condensation as well as DNA fragmentation. In a preliminary study, irradiation (0.5 Gy) of mice thymocytes showed detection of apoptosis using double fluorescence labeling of cells by annexin V-FITC and propidium iodide (Collaborative work with Prof. T. Nomura, Japan 2000 Unpublished results). Inactivation of membrane receptors has been shown in irradiated human B-lymphocytes [30]. Involvement of membrane lesions in cell killing by apoptosis and contributions of membrane proteins in the promotion of radio-resistance are some of the exciting areas for radiobiological investigations.

Signaling and Apoptosis

Apoptosis or program cell death is initiated by a variety of external and internal stimulus such as growth factor withdrawal, metabolic or cell-cycle alterations, activation of death receptors on cell surface and DNA damage. When a cell is exposed to ionizing radiation at least two major targets, membrane and DNA, generate signal cascade (although signals may also originate in the cytoplasm). Each of these targets initiates a signal that is ultimately translated into either non-death stress response or a death response. The signals originated from these locations may follow independent pathways with or without common elements. The functioning of network of signals is complicated as some increase while others decrease. However, it is yet unclear how radiation triggers apoptosis. Experimental data suggest that apoptosis mediated by DNA damage occurs by p53 dependent and independent mechanisms. However, both DNA and membrane initiated apoptosis are inhibited by bcl 2, a death suppressor. In response to stress, the ratio of death suppressor and death promoter determine inhibition or stimulation of apoptotic cell death. FAS-legend pathway of apoptosis observed in T and B cells involves binding to a death domain on membrane surface initiating activation of a cascade of proteases. Apoptosis induced by TNF appears to follow the pathway similar to FAS Recent investigation has shown the involvement of cytoplasmic caspases in induction and execution of apoptotic process. There is a growing realization that rational approach for drug design based on gene or signal transduction pathway might prove preferable to random screening of compounds for therapeutic goals.

Modification of Radiation Damage

Modification of radiosensitivity of cell is evidently of immediate importance to radioprotection and radiotherapy. A number of cellular factors modify free radical and radiation mediated damage of cell. Some enzymes and reductants involved in cellular antioxidant defense (catalase, peroxidase, SOD, glutathione, vitamin C and E) act as radioprotective agent by virtue of their reaction with various forms of reactive oxygen species or their reaction products ($^.OH$, superoxide, singlet oxygen, alkoxyl and peroxy radicals, peroxides and hydroperoxides, and aldehydes). Scavenging of hydroxyl radicals prevent or reduce free radical mediated reactions but other forms of reactive oxygen ($O^._2$, $LO^._2$ etc.) are involved directly or indirectly in radiation damage [9]. Irradiation of cells in oxygenated condition renders them 2-3 times more sensitive to radiation killing. The ratio of radiation dose given under anoxic condition to produce a given effect and dose given under fully oxygenated condition known as oxygen enhancement ratio (OER) gives a quantitative measure of sensitizing effect. Certain chemicals and drugs have been found to possess ability to mimic oxygen effect (radiosensitisers). Electron affinic compounds, namely, NEM, iodoacetamide, TAN, misonidazole, metronidazole and DNA analogs like halogenated nucleotides have been extensively investigated for their ability to enhance cellular radiosensitivity. In addition, some membrane specific drugs, namely, procain, tertracaine and chlorpromazine have been shown to modify radiosensitivity of bacterial and mammalian cells [20]. The cellular radiation damage is also markedly modified by natural free radical scavengers like α-tocopherol and carotenoids or by membrane structural ordering agents like cholesterol [32]. Studies have demonstrated that inclusion of cholesterol in membrane of egg yolk lecithin liposomal membrane inhibit the radiation induced oxidative damage which was explained on the basis of rigidization effect of cholesterol in biological membrane [32]. Sensitivity of cells to radiation is known to depend on the stage of cell cycle as well as on damage repair capability. It is commonly understood that the cell or tissue with high proliferative rate and low repair efficiency exhibit greater radiosensitivity but recent observations point to a change in this perception.

A major limitation of radiotherapy is the non-selective killing of tumor and normal cells. Since the boundary of tumor is not well defined from normal cells, the radiation treatment of tumor causes lethal damage to them producing severe side-effects putting restriction in the treatment. Recent technological developments in imaging techniques have made significant improvement in localized tumor treatment but considerable scope of radiobiological research for selective killing of tumor cells while sparing normal cells. Radioresistant hypoxic tumor cells pose a serious problem to radiation therapy. In growing tumor (a few millimeter diameter) the core of tumor mass becomes partially or completely devoid of oxygen generating a hypoxic or an anoxic region. The hypoxic region of tumor is known to display radioresistance and eventually failure of cancer radiotherapy in clinic. Extensive studies have been carried out to overcome the hypoxia problem by developing radiation hypoxic sensitizers [20, 39]. Fractionation of dose applicable for radiotherapy i.e. small radiation doses given at daily intervals, produced less damage to normal cells than to the tumor. Presumably, normal cells may recover from the damage more efficiently than tumor cells. Fractionated procedure of radiotherapy is a significant contribution of radiobiology. The fractionated course of radiation therapy produced definite therapeutic advantage and it has become a routine procedure in radiation treatment of cancer patient.

Chemotherapy of cancer is of limited value due to chemoresistance of tumor and unacceptable toxicity of antitumor drugs. Radiotherapy combined with chemotherapy is a often practiced protocol. Since hypoxic cells are resistant to radiation, some reduction of tumor mass by drugs may improve the blood supply producing reoxygenation of tumor and increase in radiosensitivity [31]. Many new approaches of combinational therapy are currently under development for achieving effective therapeutic advantage. Liposomes are known as one of the efficient mode of drug delivery system for increasing drug toxicity to tumor. Liposomal mode of anti-cancer drug delivery has already shown considerable success in improving the treatment outcome when combined with radiotherapy. A notable demonstration has been made in our laboratory with experiments conducted on transplanted mouse tumor after administration of thermosensitive liposomes (TSLs) encapsulated with doxorubicin which effectively reduced the tumor growth time in locally heat (42-43 °C) treated fibrosarcoma grown on the hind leg of mice [8]. Another rapidly growing and promising approach for overcoming resistance to drug or radiation is by use of combinational modalities based on electroporation technology [21, 17, 27, 28–29, 24]. Combination of electroporation and chemotherapy has already shown response *in vitro* studies [24], which is expected to meet the challenge of overcoming the chemoresistance of tumor cells. *In vitro* investigations on Ehrlich ascites tumor cells treated in combination with radiation and electroporation have showed synergistic effects on cell killing [22, 37, 24]. Further work may provide a new mode of cancer radiotherapy treatment (Electrochemoradiotherapy).

Radiation Biology and Cancer Radiotherapy

Basic research in radiation biology is valuable and constitutes major resource for innovative ideas but need for providing a more effective treatment for cancer patients is of paramount importance. However, basic research needs to be translated into radiotherapy practice in clinic. New therapies exploiting differences in properties of tumor and normal cells are worthy of investigation and it holds a promise for therapeutic gain. Fractionation of radiation dose will remain a major endeavor for further development of improved treatment protocol.

Apoptosis pathway is disrupted in tumor cell conferring resistance to antitumor therapies. Failure or poor response of tumor cell to radiation therapy reflects lack of susceptibility to induction of apoptosis. Therefore, understanding the mechanism of radiation induced apoptosis holds enormous promise to improve cancer radiotherapy. It seems possible to exploit intrinsic apoptosis threshold differences in normal and tumor cells to achieve a long-term goal of selectively killing of tumor cells. Apoptotic potential may prove a regulator of both intrinsic and extrinsic determinant tumor response to radiation therapy. There seems a relation between apoptosis susceptibility and cellular radiosensitivity, which needs to be exploited for translating this knowledge to treatment outcome in clinic.

Genetic mutation rendering resistance to cells against apoptosis not only produced malignant transformation but also protected them from anti-cancer agents. Currently it is believed that apoptotic resistance of tumor cell confers resistance to radiation and chemotherapy. The involvement of intracellular factors like Bcl-2, Myc, p53 and, other oncogene and tumor suppresser genes are known in radiation induced apoptotic death [2, 3]. In certain types of tumor there is a good correlation between the loss of p53 dependent apoptosis and poor prognosis. However, this does not seem true in certain other types of tumors such as Wilms tumors, testicular carcinoma and acute lymphoblastic leukemia. Insensitivity to apoptosis induction seems to be the mechanism of failure of antitumor treatments and modulating the trigger of apoptosis in tumor may offer effective treatment strategy for cancer. Multitude of factors like stage of cell (cell cycle), nature of radiation (LET), oxygen concentration and total dose and dose rate determine the apoptotic process in cells following radiation exposure. Studies on molecular pathways of programmed cell death have provided new opportunities in cancer biology but extensive work is warranted to harness the inherent power of cellular apoptosis for effective therapeutic gain.

Radiobiological research and society

Nuclear technology and procedures have become an essential part of human life. Increasing use of nuclear energy for generation of electricity involves production of ionizing radiation, which has potential to adversely affect humans. Medical radiation biologists have an important role to play in understanding the principles and concepts of radiation action on humans in order to guide and advise the practicing physicians dealing with the situation more scientifically and effectively. During emergency in a nuclear reactor or in situations of military weapon program, a physician trained in radiobiology will be required to find out the dose of radiation released who should be able to provide appropriate advice and adequate information to public and patients. A radiation biology trained scientist may give aid and advice to radiation emergency management on the basis of a rational approach. A scientist/physician is viewed by society to possess knowledge relevant to decisions concerning health risks. There is an unique role of informed radiobiologists to address issues raised by the public and social bodies and to remove misinformation which may cause unnecessary apprehension and avoidable fear in the mind of public. Therefore, radiation biology trained professionals have an important role together with various emergency management agencies, namely, media, police and fire personnel including feeding scientific information to governmental administrative offices responsible to public health and property. An updated knowledge in developments in radiation science is evidently of considerable interest to the modern society.

New Prospects and Future Challenges

Basic studies on the molecular mechanism of radiation action on cells and their modification by physical and chemical agents form a frontier area of research because of its impact on both radiation protection and improvement of radiotherapy. Radiation biology has to provide unequivocal answer to several paradoxical and controversial issues such as cellular radiosensitivity, adaptive response phenomenon, low dose radiation effects, contributions of DNA and membrane damage to cell death. Nuclear technologies will grow in the field of energy and environment, medicine and agriculture, health care and industry. Nuclear medicine methods will expand in sophistication and refinement allowing detection, imaging and examination of pathological tissues which are impossible without it. Prospects of basic radiobiology applied to cancer radiotherapy either alone or in combination with other developing modalities of treatment, namely, ECT, chemotherapy, hyperthermia, and liposomes are enormous. Role of basic research to understand mechanisms of biological effects of radiation will help optimize present radiation related technologies and may expand new applications in diverse areas such as food irradiation, mutation breeding and crop improvement, sewage treatment and radioprotection devices.

Apoptosis, in particular, has to play an important role in carcinogenesis and radiotherapy [2]. Focus of studies on the mechanism of cell death will dominate because of a growing realization that tumor growth is not just a function of cell proliferation but also of cell death. Tumor radiobiology is anticipated to witness a sea change in perception and approach. For instance, intrinsic killing power of drugs or radiation may not be as important as their ability to persuade cancer cells to initiate apoptotic death program to destroy themselves. Therefore, search for modulators (initiators and inhibitors) of apoptotic cell death is highly warranted. The greater challenge faced is to effectively translate radiobiological research into radiotherapeutic practice. Striking connections have been found among p53, oncogenes and apoptosis but apoptosis has been observed to proceed in mutated p53 or overexpressed Bcl2 and plausible explanations are presently lacking. Revolutionary progress may be hoped in cancer therapy when strategy can be developed to selectively induce death program in tumor cells while sparing normal cells to overcome present limitations of antineoplastic therapies, namely, radio and chemotherapy. The differential apoptotic index may provide a prognostic indication of tumor in response to radiotherapy. The challenge of harnessing the power of apoptotic response to clinical gain continues to engage attention of scientists. However, it is hoped that these and other new approaches of improving the radiation treatment of cancer would reach a level of development to find a place in future clinical practice.

Research in radiation membrane biology needs to be focused on contributions of free radical mediated oxidative membrane damage in the mechanism of cell death. It is seen that at low dose and low dose rate membrane related functions are more sensitively affected which need to be detected and quantified. Incorporation or decorporation of radionuclides especially heavy metal radionucildes may be influenced by membrane modification and by using artificial liposomal membrane as carriers for chelating agents. Radiation induced perturbations on membrane may be prevented or modulated by certain membrane specific drugs and other agents opening a new perspective in membrane-based radioprotection and radiosensitization. Radiation biology of membrane might open a new direction for biological indication of radiation injury especially at therapeutic and low doses of radiation. The radiation protection for occupational, medical and accidental radiation exposed persons seems to be one of the major challenges of radiation biology for new the millennium.

Acknowledgements

My thanks are due to all my students, colleagues and collaborators who provided support during these studies. I especially acknowledge the help and assistance of my younger colleague, B.N. Pandey, who put sincere efforts in organizing the material for this presentation.

References

1. Alper, T. (1977) In *Membrane Toxicology*, Eds. M.W. Miller and A.E. Shamou NY: Plenum Press. pp. 139.
2. Blank, K.R., M.S. Rudoltz, G.D. Kao, R.J. Muschel and W.G. McKenna (1997) *Intl. J. Radiat. Biol.* 5: 455.
3. Brown, J.M. and Wouters, B.G. (1999) *Cancer Res.* 59: 1391.
4. Coggle, C.E. (1983) *Biological Effects of Radiation*, 2nd Ed., Taylor and Francis Ltd., London.
5. Cregen, S.P., B.P. Smith, D.L. Brown and R.E.J. Mitchel (1999) *Intl. J. Radiat. Biol.* 75: 1069.
6. Curie, M.S. (1904) Reprinted from Chemical News, D.Van Nostrand Co, New York.
7. Gillies, N.E. (1997) *Intl. J. Radiat. Biol.* 71: 643.
8. Gopal, A.R. (1996) Ph.D. Thesis, Mumbai University, Mumbai.
9. Greenstock, C.L. (1988) *Pharmac. Ther.* 39: 139.
10. Haimovitz-Friedman, A., C.C. Kan, D. Ehleiter, R.S. Persaud, M. McLoughlin, Z. Fucks, and R.N. Kolesnick (1994) *J. Expt. Medi.* 180: 525.
11. Halliwell, B. (1994) *Lancet*, 344: 721.
12. Hewitt, H.B. and C.W. Wilson (1959) *Nature*, 183: 1060.
13. Kerr, J.F., A.H. Wyllie, A.R. Currrie (1972) *Br. J. Cancer*, 26: 239.
14. Koteles, G.J. (1979) *Atomic Energy Review* 171: 1.
15. Koteles, G.J. (1982) *Radiat. Envirn. Biophys.* 21: 1.
16. Mishra, K.P. (1982) *Ind. J. Biochem. Biophys.* 19: 21.
17. Mishra, K.P. and B.B. Singh (1986) *Ind. J. Expt. Biol.* 24: 537.
18. Mishra, K.P., B.B. Singh, and A.R. Gopal-Ayenger (1973) *Intl. J. Radiat. Biol.* 24: 537.
19. Mishra, K.P., Y. Nosaka, K. Akasaka, H. Hatano, and C. Nagata (1978) *Biochim. Biophys. Acta* 520: 679.
20. Mishra, K.P., V.T. Srinivasan, and B.B. Singh (1980) *Radiat. Res.* 8: 413.
21. Mishra, K.P., V.W. Bedekar, and B.B. Singh (1983) *Ind. J. Expt. Biol.* 21: 641.
22. Mishra, K.P., M.C. Patel, R.D. Ganatra, and B.B. Singh (1985) *Bibl. Heamatol.*, 51: 115.
23. Mishra, K.P., D.C. Joshua, and C.R. Bhatia (1987) *Plant Sci.*, 52: 135.
24. Mishra, K.P., H.D. Sarma, G.S. Nanda, and B.B. Singh (1996) *Progr. Mol. Biol.Biophys.* 65: 98.
25. Moorthy, P.N., K.N. Rao, K.P. Mishra and B.B. Singh (1972a) *Radiochem. Radioanal Lett.* 10: 45.
26. Moorthy, P.N., K.N. Rao, K.P. Mishra and B.B. Singh (1972b) *Radiat. Nucl. Chem.* 1: 311.
27. Nanda, G.S. and K.P. Mishra (1993) *Bull. Electrochem.* 9: 547.
28. Nanda, G.S. and K.P. Mishra (1994a) *Bioelectrochem. Bioenerget.* 34: 129.
29. Nanda, G.S. and K.P. Mishra (1994b) *Bioelectrochem. Bioenerget.* 34: 189.
30. Ojeda, F., H.A. Diehl and H. Folch (1994) *Scann. Microsc.* 3: 645.
31. Okawa, T. and M. Kita-okawa (1996) *Intl. J. Clin. Oncol.* 1: 3.
32. Pandey, B.N. and K.P. Mishra (1999) *Radiat. Phys. Chem.* 54: 481.
33. Pandey, B.N. and K.P. Mishra (2000a) *Appl Magn. Reson.* 18: 438.
34. Pandey, B.N. and K.P. Mishra (2000b) *Bull. Res. In Press.*
35. Puck, T.T. and P.Z. Marcus (1956) *J. Expt. Med.* 103: 653.
36. Rols, M.P., F. Dahhou, K.P. Mishra and J. Teissie (1990) *Biochem.* 29: 2960.
37. Ramakrishnan, N., D.E. McClain and G.N. Catravas (1993) *Intl. J. Radait. Biol.* 63: 693.

38. Shenoy, M.A., K.P. Mishra, B.B. Singh and A.R. Gopal-Ayenger (1974) *Intl. J. Radiat. Biol.* 25: 303.
39. Singh, B.B. (1980) *Atomic Energy Review* 181: 171.
40. Stark, G. (1991) *Biophy. Biochim. Acta*, 1071: 103.

Radiobiology and Bio-Medical Research
Edited by K.P. Mishra
Copyright © 2004 Narosa Publishing House, New Delhi, India

7. Molecular Radiobiology: Plasmid pMTa4 as a Tool for Studying Effects of *g*-Radiation *in vitro* and *in vivo*

J.O. Humtsoe and R.N. Sharan*

Radiation & Molecular Biology Unit, Department of Biochemistry,
North-Eastern Hill University, Shillong 793 022, India

Abstract   Radiation induced damage to DNA molecule is known to be critical. Base or nucleotide damage, strand breaks and clusters of damages are some of the important lesions produced by radiation in a DNA molecule. However, molecular intricacies of such events are not clearly understood obscuring molecular mechanisms of radiation-induced damages and their repair. A well-defined plasmid DNA construct, pMTa4, along with selected restriction endonucleases have been utilized in this line of study to elucidate some events that follow irradiation *in vitro* and *in vivo*. The results reveal that GC-rich nucleotide sequences of pMTa4 were preferably affected by ^{60}Co γ-irradiation *in vitro* generating premutagenic lesions in a non-random way. The results also show that even under repair non-permissive conditions repair activities went on *in vivo*, albeit with very low fidelity causing misrepair. Under repair permissive conditions, in contrast, high fidelity repair was observed. These observations are likely to have significant bearing on our present understanding of inherent radiosensitivity and genome instability. Use of a simple plasmid DNA molecule can potentially provide clear insight into molecular consequences of irradiation on DNA and should be exploited further.

Introduction

Plasmid DNA, besides being the foundation of biotechnology, can be a potentially very useful tool for studying radiation-induced DNA damages. Its simple form, small size and known nucleotide (NT) sequence provides a convenient possibility of qualitatively and quantitatively determining the DNA lesions. In conjuncture with endonucleases, especially restriction endonucleases (RE), plasmid DNA can become an extremely powerful tool in analyzing the manifestation of DNA damage following irradiation. In recent past, this approach has been employed in some such studies of radiation-induced DNA damages. Limited studies using plasmids have provided new insights in studies of repair, mutation induction, alterations of DNA structures by alkylation, drug-DNA-interaction and in measurement of DNA strand breaks [19, 1, 22, 9]. RE is highly NT specific in its action and cleaves DNA only if is not modified at its restriction site. Therefore, by analyzing the fragments of a plasmid produced by a RE, one can not only deduce the order of the segments within the original DNA molecule, but also possibly study the alterations or modifications of the NT sequence that might have been induced by radiation. Further, generation of single strand breaks (SSB) and double stranded breaks (DSB) in a DNA molecule can also be very conveniently monitored and analyzed.

Base or NT damage is one of several distinct lesions produced by radiation in a DNA molecule. It involves chemical alterations of the bases that can result from severance of a side group from the base or damage to the ring structure itself. A large number of base products have been identified and reported in literature [33]. Ionizing radiation has also been shown to induce clustered DNA damage containing oxidized purines and pyrimidines and abasic sites [29]. In recent years, studies on identification of such damage have been carried out to define specific molecular damages in an attempt to understand the non-random biological response to radiation. In these investigations repair enzymes like formamidopyrimidine-DNA glycosylase or endonuclease III have been used. These enzymes were found to be useful in characterization of DNA damage induced by reactive oxygen species and photosensitizers in cellular and cell-free systems [7] besides by γ-radiation [16]. The damages usually get converted to SSB, which can be then measured by several methods like nick translation [5], gel electrophoresis [21], or comet assay [26]. The use of RE offer several additional advantages due to its very high NT specificity and could be a potential tool for elucidation of molecular mechanism of radiation induced DNA damage [12].

Using a well-defined plasmid construct, designated as pMTa4, and selected RE, this piece of work was designed in an attempt to study influence of NT sequence on DNA damage induced by γ-radiation and its repair in a cellular (*in vivo*) as well as an acellular (*in vitro*) condition.

Materials and Methods

E. coli culture and preparation of plasmid DNA:
XL 1 strain of *E. coli* bacterium harboring a 6173 bp plasmid construct, designated as pMTa4 (kindly provided by Prof. C.H. Schroeder, DKFZ, Heidelberg), was grown at 37 °C in LB medium containing ampicillin (50 μg ml^{-1}). After overnight culture, the plasmids were isolated by High Pure Plasmid Isolation kit (Boehringer Mannheim, Germany). The yield and purity of the isolate were determined by measuring the absorption ratio A_{260}: A_{280}.

Irradiation source and experimental protocols:
E. coli culture (2 ml of overnight culture) on ice or an aqueous solution of isolated pMTa4 (41.25 μg ml^{-1}) at room temperature was γ-irradiated in a ^{60}Co Gamma chamber 900 (~0.2 Gy sec^{-1}) to accumulate doses of 10, 20 and 30 Gy (cellular or *in vivo* condition) or 30, 60, 120 and 240 Gy (acellular or *in vitro* condition), respectively. The irradiated samples were stored on ice until analysis. From the irradiated *E. coli*, plasmid DNA was isolated similarly either immediately after irradiation (repair non-permissive) or after a post-irradiation repair incubation

of 60 minutes at 37 °C (repair permissive). The plasmid preparations were analyzed further.

Analysis of plasmid DNA

Irradiated and non-irradiated pMTa4 samples (~1 μg) were run on a one percent agarose gel electrophoresis using TAE buffer for monitoring the SSB and DSB induced by irradiation. In addition, the pMTa4 preparations were fragmented with different RE such as, *ACC I Hae II, Ksp I, Nci I, Bgl II, Dra I, Pvu II, Bgl I* and *Hinf I*, under appropriate conditions (Boehringer Mannheim, Germany), and analyzed by gel electrophoresis. Ethidium bromide stained bands on the gels were seen through an UV transilluminator and scanned (GS-690 Imaging Densitometer).

Strand break analysis and determination of fragment sizes

The relative amounts of DNA in each of the plasmid topological form resulting from irradiation were determined by scanning. The sum of the pixel intensities of the DNA bands were calculated using Molecular Analysts™/PC Windows software, version 1.5. The approximate sizes of the DNA fragments generated by RE were determined using the same software. Based on the Rf values of the different fragments, their molecular weights (sizes; bp) were obtained against known λ DNA double digest marker through third order cubic regression.

Results

Effect of γ-radiation on pMTa4 *in vitro:*

Figure 1 depicts the gel photograph and plot revealing the changes in the three

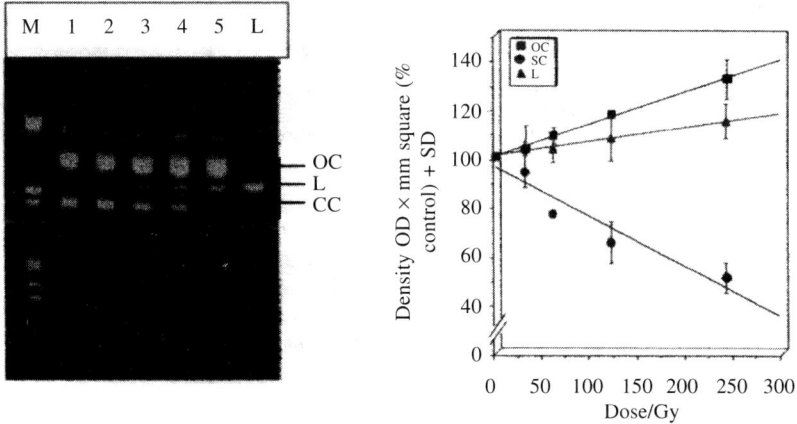

Figure 1. Agarose gel electropherogram of pMTa4 showing its supercoiled (CC), open circle (OC) and linear (L) topological forms (left). Lane 1 shows the control plasmid (unirradiated) while lanes 2 to 5 show *in vitro* γ-irradiated plasmid with 30, 60, 120 and 240 Gy, respectively. M represents λ DNA double digest marker while L represents *Nco I* linearized non-irradiated pMTa4. Densitometric analysis of 3 topological forms of pMTa4 is shown in the right panel; data points are from two independent experimental sets and represent the mean $\pm$ SD.

topological forms of pMTa4 in its non-irradiated (control) state and that follow irradiation. It indicates dose dependent reduction in the quantity of supercoiled (SC) form, while dose dependent increase in the quantities of open circular (OC) and linear (L) forms were observed.

Effect of γ-radiation on the pMTa4 restriction sequences for different RE

Table I lists different restriction endonucleases used in this study along with their restriction sequence and NT sequence at fragment-end in pMTa4 plasmid. Figure 2 shows the restriction fragmentation patterns of non-irradiated (control) pMTa4 generated by *Hae II* and *Nci I*, and that after exposure of the plasmid to varying doses of γ-radiation. The irradiated plasmid restricted with *Hae II* and *Nci I* showed additional slow migrating bands or large DNA pieces on the gel which were entirely absent in the non-irradiated control. The formation of the extra fragments (Fig. 2A lanes 2 through 5) remained unchanged even at the maximum dose of 240 Gy for *Hae II*. The *Nci I* restriction showed formation of four distinct additional bands (Fig. 2B). The band intensities increased with radiation dose exhibiting dose dependence (Table 2), as revealed by densitometric quantification. In contrast, the fragmentation pattern of pMTa4 generated by other RE did not show any difference between irradiated and non-irradiated DNA samples (results not shown).

Table 1. Characteristics of the different restriction endonucleases (RE): RE were selected depending on their restriction sites: blunt- and staggered-ends as well as their NT composition in the site.

Restriction endonuclease (RE)	Restriction site	Fragment-end generated by RE cleavage	Effect on the restriction following radiation
Acc I	–GC/(A,C) (T,G)AC–	–GC	–
	–CA(T,G)(A,C)/TG–	–CA(T,G)(A,C)	
Bgl I	–GCCGTTT/GGGC–	–GCCGTTT	–
	–CGGC/AAACCCG–	–CGGC	–
Bgl II	–A/GATCT–	–A	–
	–TCTAG/A–	--TCTAG	
Dra I	–TTT/AAA–	–TTT	–
	–AAA/TTT–	–AAA	
Hae II	–GGCGC/(T,C)–	–GGCGC	+
	–C/CGCG(A,G)–	–C	
Hinf I	–G/ANTC–	–G	–
	–CTNA/G–	–CTNA	
Ksp I	–CCGC/GG–	–CCGC	–
	–GG/CGCC–	–GG	
Nci I	–CC/(C,G)GG–	–CC	+
	–GG(G,C)/CC–	–GG(G,C)	
Pvu II	–CAG/CTG–	–CAG	–
	–GTC/GAC–	–GTC	

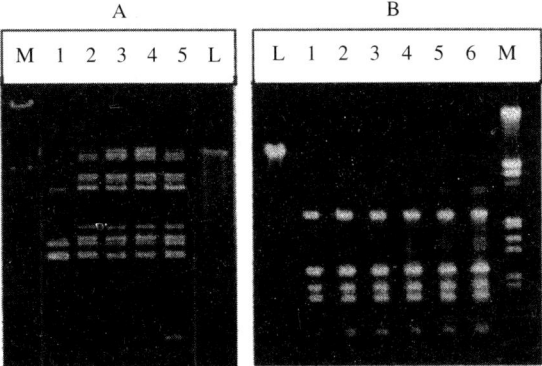

Figure 2. Agarose gel electropherogram of pMTa4 restricted with *Hae II* (A) and *Nci I* (B). Restriction with *Hae II* was carried out after 0 (control), 30, 60, 120 and 240 Gy of *in vitro* γ-radiation (lanes 1 through 5), respectively. Restriction with *Nci I* was done following 0 (control), 15, 30, 60, 120 and 240 Gy of *in vitro* γ-radiation (lanes 1 through 6), respectively. M represents λ DNA double digest marker while L represents *Nco I* linearized non-irradiated pMTa4.

Table 2. Densitometric quantification of the four extra fragments produced by *Nci I* after γ-irradiation (see Fig. 2B). The four fragments in order of decreasing size are denoted as F1, F2, F3 and F4. Data points are from a single experimental set.

Dose (Gy)	Density OD x mm^{-1} (% control)			
	F1	F2	F3	F4
0 (control)	100	100	100	100
15	193.85	186.49	181.14	173.33
30	228.07	213.79	206.20	200.74
60	230.99	242.52	231.26	225.18
120	260.23	254.59	241.43	234.81
240	296.78	281.60	294.54	288.64

Analysis of *Hae II* and *Nci I* generated fragmentation pattern after γ-irradiation

Table 3 shows the sizes of *Hae II* and *Nci I* generated extra fragment that were calculated using MA software in comparison to the theoretical sizes that should be generated. Out of the several possible theoretical combinations of fragments generated by *Hae II*, the values falling nearest to the observed sizes calculated by software were taken as their actual fragment size (Table 3, A). An approach similar to that of the *Hae II* fragment analysis was employed in the identification of the extra fragments produced by for *Nci* (Table 3, B).

Effect of γ-radiation on the pMTa4 *in vivo*

The results of analysis of pMTa4 isolated from irradiated *E. coli* cells (without and with post-irradiation repair incubation) are shown in Figure 3. In absence of

Table 3. **pMTa4 size analysis of the extra fragments generated by *Hae II* (panel A) and *Nci I* (panel B) restriction after *in vitro* γ-irradiation. Left columns represent the values obtained from the molecular analyst software, while right column represent the values of predicted or expected fragments generated by possible combinations that would result due to incomplete digestion.**

A		B	
Approximate fragment Size (bp) calculated by MA software γ-radiation	Predicted actualy size in bp with their possible fragment comb. inations	Approximate fragment size (bp) calculated MA software γ-radiation	Predicted actual size in bp with their possible fragment combinations
6356	6173	3208	3272
6055	5802	2966	2901
5605	5589	1537	1514
4300	4234/4326	1387	1387
4126	3864/3856		
3721	3539		
2462	2588		
2261	2316/2303		

repair incubation, both OC (representing SSB) and L (representing DSB) forms of pMTa4 increased in a radiation dose dependent manner (Fig. 3 – minus). In addition, progressively slow migrating bands of DNA were observed between OC and L bands generating a 'ladder' (Fig. 3) The intensity of bands in the 'ladder' showed dose dependence. Both SSB and DSB were efficiently repaired upon repair incubation accompanied by almost total fading of the 'ladder' (Fig. 3–plus) The plasmid did not show any difference in restriction patterns by *Hae II* and *Nci I* before and after irradiation (results not shown).

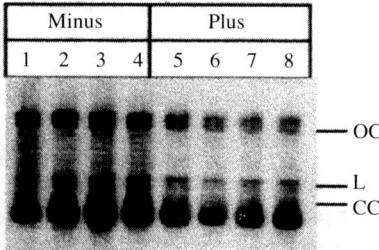

Figure 3. **Agarose gel electropherogram of pMTa4 isolated from *E. coli* (*in vivo*) following irradiation accompanied with repair non-permissive (minus; lanes 1 through 4)) and permissive (plus; lanes 5 through 8) conditions. Lanes 1 and 5 show pMTa4 isolated from unirradiated *E. coli* cells while lanes 2 to 4 and 6 to 8 show the pMTa4 following 10, 20 and 30 Gy irradiation, respectively.**

Discussion

The four NT of DNA molecule (G, A, T and C) are constant targets of damage by chemical processes such as methylation, depurination, deamination and oxidation. Induction of DNA damage by radiation is mostly caused by oxidation via ˙OH mediated indirect effect [33]. The 5-C and 6-C of the pyrimidines and the 4-C and 8-C of purines are the most probable sites known for the attack resulting into a plethora of modified bases [8]. Such damages induce predominantly SSB. Fig. 1 shows the changes in the pMTa4 after exposure to γ-radiation representing SSB and DSB. The increase in the OC form at higher doses of radiation is an indicator of induction of SSB. ˙OH has been reported to be much more effective in the production of SSB than ˙H and e_{aq}^- that are formed when DNA is γ-irradiated in an aqueous solution [16]. Since in this study pMTa4 was irradiated in an aqueous solution (*in vitro*), it will be reasonable to assume that induction of SSB was primarily due to the water derived ˙OH and to a lesser extent ˙H [16], and some direct effect. This is in good agreement with data on dose dependent response to radiation [20, 32]. On the other hand, in cellular condition (*in vivo*), radiation induced significant increase in quantities of both OC and L forms suggesting that both SSB and DSB were being induced in pMTa4 when *E. coli* cells were γ-irradiated (Fig. 3-minus). When the irradiated *E. coli* cells were incubated at 37 °C for 60 minutes permitting repair prior to pMTa4 isolation, there was significant drop in both OC and L forms suggesting that most damages could be repaired in this situation (Fig. 3–plus). It has to be noted that aqueous pMTa4 (*in vitro*) had no such repair possibility (Fig. 1).

The restriction capability of RE was observed to vary after *in vitro* plasmid DNA was exposed to γ-rays (Fig. 2). While efficiency of the plasmid restriction by *Acc I, Ksp I, Bgl II, Dra I, Pvu II, Bgl I* and *Hinf I*, used in this study, remained unchanged (not shown), it was significantly reduced for *Hae II* and *Nci I* (Fig. 2A and B) after γ-irradiation. Such sub-optimal restriction can result when either the restriction site is shielded by DNA binding chemical compound [9] or due to possible radiation-induced base modification [24]. The former possibility can be ruled out for the experimental set-up used in this investigation since it lacks any other component which may interact with the plasmid DNA in an aqueous solution. The partial restriction by *Hae II* appears to be dose independent as the intensities of fragments generated after the lowest dose of 30 Gy did not change even after the dose of 240 Gy (Fig. 2A, lanes 2 through 5). Result from *Nci I* restriction (Fig. 2B; lanes 2 through 6) also followed a similar pattern. The increase in the intensity of the additional fragments (Table 2) may, on the other hand, indicate a dose dependent manifestation of effect. The non-random generation of fragments due to sub-optimal restriction suggest that all restriction site for *Hae II* and *Nci I* in pMTa4 were not equally susceptible to the effect by radiation. For such sub-optimal restriction to occur, it is likely that pMTa4 underwent certain specific alteration or modifcation in selected *Hae II* and *Nci I* restriction sites such that only these sites became resistant to RE. This assumption logically requires substantiation by looking for the specific NT composition of the restriction sites for *Hae II* and *Nci I* in order to explore the possibilities of identifying changes that may be responsible for such partial fragmentation. A closer look into *Hae II* and *Nci I* restriction site revealed its GC-richness (Table I). Unlike the others, except *Ksp I*, it was notable that *Hae II* and *Nci I* produced 100 percent GC staggered-ended DNA pieces (Table I). The other RE that generated either staggered- or bluntended pieces did not show the GC-motif feature. The NT composition in the flanking region around the restriction sequence of *Hae II* and *Nci I* too indicated high GC content [13].

Although the type of modified base was not determined in this study, a wide range of radiation-induced bases damages are known [30]. Of the several types, 8-oxoguanine (OG) has been identified as the major product [15]. Fuciarelli *et al.* [10] found that when DNA was γ-irradiated in oxic condition, high amount of OG was formed. Under conditions where the ·OH are the only damaging species, formamidopyrimidine glycosylase sensitive sites were formed in high amount [16] indicating a high formation of guanine product. Deamination or demethylation of C nucleotide is also known to occur following irradiation. Since the NT sequences effected are GC-rich, the resistance to restriction after irradiation may be due to modification of G and/or C nucleotides. Regardless of the type and nature of modification, the observations indicate that GC-rich motif was more frequently modified or damaged upon interaction with radiation.

In the cellular (*in vivo*) situation, there was a dose dependent increase in OC as well as L forms of pMTa4 after irradiation and there was apparent repair of most strand breaks under post-irradiation repair permissive condition (Fig. 3). Under repair non-permissive condition, it also showed a 'ladder' due to progressively slow migrating DNA bands between OC and L bands, which faded significantly upon repair incubation. Further, the plasmid DNA did not show any resistance to *Hae II* and *Nci I* restriction (not shown) in sharp contrast to the *in vitro* results (Fig. 2). These results suggest some very interesting insight into induction of radiation damage to pMTa4 DNA. At first, partial repair activity must have been on even under repair non-permissive condition (on ice), which could repair, at least, the base modifications efficiently since no resistance to RE restriction was observed *in vivo*. Secondly and as expected, under repair permissive condition, both SSB and DSB were repaired which was not the case in absence of repair incubation (Fig. 1). Thirdly, there was formation of a 'ladder' in repair non-permissive condition, which faded significantly upon repair incubation.

The cause of 'ladder' formation needs further explanation. It is known that plasmid DNA in its native conformation is supercoiled (CC form) and migrates fast on agarose gel. This band becomes significantly slow migrating upon relaxation of the conformation (OC form) caused by nicks or SSB, which reduces its extent of supercoiling. The L form, caused by DSB, migrates in between the two bands (see Fig. 1). The 'ladders' represents a collection of plasmid DNA molecules with intermediate states of supercoiling resulting in progressively slow migrating bands on agarose gel. Since there are indications of partial repair activity under repair non-permissive condition, it can be assumed that some of the radiation-induced nicks or SSB were also being repaired. However, the fidelity of this repair seemed to be poor, falling short of complete repair. This may cause wrong joining of DNA strands (or misrepair) creating different degrees of supercoiling in different pMTa4 molecules. This could lead to collection of pMTa4 molecules with different degrees of supercoiling and produce the 'ladder'. Apparently some critical components of repair machinery were absent in repair non-permissive condition. Upon repair incubation, the critical components also became available, perhaps due to induction, leading to high fidelity of repair exhibited by fading of the 'ladder'.

While confirmation of this hypothesis may involve some additional work, there are indirect supports available in literature. Large numbers of studies have reported high mutagenecity at GC rather than AT [23, 3]. Mutation studies using MSH2 deficient mice have indicated a greater percent of mutation in GC than AT [25, 27]. Chemically modified G or C causes mispairing, where an error-prone polymerase frequently introduces a wrong base opposite these NT. OG preferentially pairs with A and causes G to T transversion [14]. 5-methyl C in a CpG motif can undergo deamination to produce C to T transition [26]. C could also spontaneously deaminate to produce uracil [17] or be oxidized to 5-hydroxycytidine [31] leading to a C to T transition. With such evidences of GC-vulnerability to radiation, it is likely that radiomodified GC nucleotides would form important premutagenic lesions. This indication also points that clusters of GC in the DNA molecule may very likely form hotspots for radiation-induced damages. While further detailed investigation would be required, it opens up a likely possibility that inherent radiosensitivity and genome instability may be at least partly determined by the GC-richness of NT sequence in the DNA. Similarly, GC-rich sequences, referred to as 'CpG islands', have been reported to occur frequently in many human genes [6, 11, 28]. The results from this piece of work suggests that most genes (rich in CpG islands), regardless of its state (transcribing or non-transcribing), might be effected by radiation with equal probability. This perhaps partially explains the high radiosensitivity of humans and current observation of induction of almost uniform damage in active or inactive genes [2, 18].

While more work, especially with mutants of repair system in *E. coli*, is currently on, it is clear that plasmid DNA has all potentials to become highly useful in studies in the domain of molecular radiobiology. The simple systems provides deep and clear insight into the molecular consequences of irradiation on DNA and should be exploited to elucidate molecular intricacies of radiation induced damages.

Acknowledgements

Part of grants for this work came as a research fellowship to JOH from UGC-NEHU. RNS is grateful to Prof. T. Nomura, Osaka, University, Japan for help in conducting a part of this work and for constructive critical comments.

References

1. Bien, M., H. Steffen and D. Schulte-Frohlinde (1988) Repair of the plasmid pBR322 damaged by gamma irradiation or by restriction endonucleases using different recombination-proficient *E. coli* strains. *Mutat. Res.* 194: 193–205.
2. Bunch, R.T., D.A. Gewirtz and L.F. Povirk (1995) Ionizing radiation-induced DNA strand breakage and rejoining in specific genomics regions as determined by an alkaline unwinding/southern blotting method. *Int. J. Radiat. Biol.* 68: 553–562.

3. Burcham, P.C. and L.A. Harkin (1999) Mutations at G:C base pairs predominate after replication of peroxyl radical-damaged pSP189 plasmids in human cells. *Mutagen.* 14: 135–140.

4. Coulondre, C., J.H. Miller, P.J. Farabaugh and W. Gilbert (1978) Molecular basis of base substitution hotspots in *Escherichia coli. Nature* 274: 775–780.

5. Czene, S. and M. Harms-Ringdahl (1995) Detection of single-strand breaks and formamidopyrimidine-DNA glycosylase sensitive sites in DNA of cultured human fibroblasts. *Mutat. Res.* 336: 235–242.

6. Dunham, I., N. Shimuzu, B.A. Roe, S. Chissoe, A.R. Hunt, J.E. Collins, R. Bruskiewich, D.M. Beare, M. Clamp, L.J. Smink, R. Ainscough, J.P. Almeida, A. Babbage, C. Bagguley, J. Bailey, K. Barlow, K.N. Bates, O. Beaseley, C.P. Bird, S. Blakey, A.M. Bridgement, D. Buck, J. Burgess, W.D. Burrill and K.P. O'Brien (1999) The DNA sequence of human chromosome 22. *Nature* 402: 489–495.

7. Epe, B., M. Pflaum, M. Haring, J. Hegler and H. Rudiger (1993) Use of repair endonucleases to characterize DNA damage induced by reactive oxygen species in cellular and cell free systems. *Toxicol. Lett.* 67: 57–72.

8. Fielden, E.M. and P. O'Neill (1991) *The Early Effects of Radiation on DNA.* 54: NATO ASI Series, Germany Springer-Verlag.

9. Financsek, I., J. Guczi and E.J. Hidvegi (1984) Interaction of GGCC sequences of DNA with anticancer dianhydrogalacticol, detected by inhibition of restriction enzyme. *Bsp I. Biochim. Biophys. Acta,* 19: 117–184.

10. Fuciarelli, A.F., B.J. Wegher, W.F. Blakely and M. Dizdaroglu (1990) Yields of radiation-induced base products in DNA: effects of DNA conformation and gassing conditions. *Int. J. Radiat. Biol.,* 58: 397–415.

11. Hattori, M., A. Fugiyama, T.D. Taylor, H. Watanabe, T. Yada, H.S. Park, A. Toyoda, K. Ishii, Y. Totoki, D.-K. Choi, E. Soeda, M. Ohki, T. Takagi, Y. Sasaki, S. Taudien, K. Blechschmidt, A. Polley, U. Menzel, J. Delabar, K. Kumpf, R. Lehman, D. Patterson, K. Reichwald, A. Rump, M. Schillhabel, A. Schudy, W. Zimmermann, A. Rosenthal, J. Kudoh, K. Shibuya, K. Kawasaki, S. Asakawa, A. Shintani, T. Sasaki, K. Nagamine, S. Mitsuyama, S.E. Antonarakis, S. Minoshima, N. Shimizu, G. Nordseik, K. Hornischer, P. Brandt, M. Scharfe, O. Schon, A. Desario, J. Reicheit, G. Kauer, H. Blocker, J. Ramser, A. Beck, S. Klages, S. Hennig, L. Reisselmann, E. Dagand, T. Haaf, S. Wehrmeyer, K. Borztm, K. Gardiner, D. Nizetic, F. Francis, H. Lehrach, R. Reinhardt and M.-L. Yaspo (2000) The DNA sequence of human chromosome 21. *Nature* 405: 311–319.

12. Humtsoe, J.O., C.H. Schroeder and R.N. Sharan (1998) [7]Li particles induced plasmid DNA damages is influenced by nucleotide sequence. *J. Radiat. Res.,* 39: 362.

13. Humtsoe, J.O. (2000) Ph.D. thesis, Study of DNA damages induced by low- and high-linear energy transfer (LET) radiation, Shillong: North-Eastern Hill University.

14. Kasai, H. and S. Nishimura (1984) Hydroxylation of deoxyguanosine at the c-8 position by ascorbic acid and other reducing agents. *Nucl. Acid Res.* 12: 2137–2145.

15. Kasai, H., H. Tanooka and S. Nishimura (1984) Formation of 8-hydroxguanine residues in DNA by x-irradiation. *GANN,* 75: 1037–1039.

16. Kuipers, G.K. and M.V.M. Lafleur (1998) Characterization of DNA damage induced by gamma-radiation-derived water radicals, using DNA repair enzymes. *Int. J. Radiat. Res.* 74: 511–519.

17. Lindahl, T. (1993) Instability and decay of the structure of DNA. *Nature* 362: 709–715.

18. Ljungman, M. (1999) Repair of radiation-induced DNA strand breaks does not occur preferentially in transcriptionally active DNA. *Radiat. Res.* 152: 444–449.

19. Lobrich, M., P.K. Cooper and B. Rydberg (1996) Non-random distribution of DNA double-strand breaks induced by particle irradiation. *Int. J. Radiat. Biol.* 70: 493–503.

20. Milligan, J.R., J.R. Aquilera and J.F. Ward (1993) Variation of single-strand break yield with scavenger concentration for plasmid DNA irradiated in aqueous solution. *Radiat. Res.* 133: 151–157.

21. Milligan, J.R., J.A. Aguilera, T-T.D. Nguyen, J.F. Ward, Y.W. Kow, B. He and R.P. Cunningham (1999) Yield of DNA strand breaks after base oxidation of plasmid DNA. *Radiat. Res.* 151: 334–342.

22. Muraiso, C., J.S. Mudgett, H. Matsudaira and G.F. Strniste (1993) A shuttle vector system for studying ionizing radiation-induced mutagenesis in mammalian cells. *J. Radiat. Res.* 34: 148–156.

23. Murata-Kamiya, N., H. Kamiya, H. Kaji and H. Kasai (1997) Glyoxal, a major product of DNA oxidation, induces mutations at G:C sites on a shuttle vector plasmid replicated in mammalian cells. *J. Nucl. Acid Res.* 25: 1897–1902.

24. Paul, C.R., E.E. Budzinski, A. Maccubbin, J.C. Wallace and H.C. Box (1990) Characterization of radiation-induced damage in (TpApCpG). *Int. J. Radiat. Biol.* 58: 759–768.

25. Phung, Q.H., D.B. Winter, A. Cranston, R.E. Tarone, V.A. Bohr, R. Fishel and P.J. Gearhart (1998) Increased hypermutation at G and C nucleotides in immunoglobulin variable genes from mice deficient in the MSH2 mismatch repair protein. *J. Expt. Med.* 187: 1745–1751.

26. Pouget, J.P., J.L. Ravanat, T. Douki, M.J. Richard and J. Cadet (1999) Measurement of DNA base damage in cells exposed to low doses of γ- radiation: Comparison between the HPLC-EC and comet assays. *Int. J. Radiat. Biol.* 75: 51–58.

27. Rada, C., M.R. Ehrenstein, M.S. Neuberger and C. Milstein (1998) Hot spot focussing of somatic hypermutation in MSH2-deficient mice suggests two stages of mutational targeting. *Immunity* 9: 135–141.

28. Reeves, R.H. (2000) Recounting a genetic story. *Nature* 405: 283–284.

29. Sutherland, B.M., P.V. Bennett, O. Sidorkina and J. Laval (2000) Clustered DNA damages induced in isolated DNA and in human cells by low doses of ionizing radiation. *Proc. Natl. Acad. Sci. USA.* 97: 103–108.

30. Wallace, S.S. (1998) Enzymatic processing of radiation-induced free radical damage in DNA. *Radiat. Res.* 150: S60–S79.

31. Wang, D. and J.M. Essigmann (1997) Kinetics of oxidized cytosine repair by endonuclease III of *Escherichia coli Biochem.* 36: 8628–8633.

32. Weinfeld, M., M.A. Chaudhary, D. D'Armours, J.D. Pelletier, G.G. Poirier, L.F. Povirk and S.P. Lees-miller (1997) Interaction of DNA-dependent protein kinase and poly (ADP-ribose) polymerase with radiation-induced DNA strand breaks. *Radiat. Res.* 148: 22–28.

33. Wiseman, H. and B. Halliwell (1996) Damage to DNA by reactive oxygen and nitrogen species: role in inflammatory disease and progression to cancer. *J. Biochem.* 313: 17–29.

Radiobiology and Bio-Medical Research
Edited by K.P. Mishra
Copyright © 2004 Narosa Publishing House, New Delhi, India

8. Peroxidation of Proteins in Biological Systems Exposed to Ionizing Radiations

Janusz M. Gebicki

Department of Biological sciences, Macquarie University, Sydney, Australia

The purpose of radiation therapy is selective killing of cells. To be fully effective, the radiologist should have the capacity to localize the treatment to the area to be sterilized, and to have the means of enhancing the effectiveness of the radiation in the critical area and of inhibiting the damage to surrounding tissues. These optimal conditions are seldom achievable in practice, partly because of significant gaps in our knowledge of the nature and relative importance of the events, which follow the deposition of the energy delivered by ionizing radiation in cells and tissues.

These events occur in a well-defined sequence, corresponding to the physical, chemical, biological and pathological stages. The sequence starts with the passage of the high-energy X or γ ray photon through the target, and the ionization and excitation of the atoms with which they interact. These, in turn, can decompose to form free radicals and excited species. We have adequate knowledge of these events, in qualitative and quantitative terms. The first point of uncertainty is the next, or the chemical stage. This is not because our knowledge of radiation chemistry of the potential biological targets of the actions of radiations is inadequate. Rather, the uncertainty lies in the identification of the *initial crucial molecular targets* damaged by the rays. These targets are defined as crucial either by their nature as *vital targets*, or by their ability to transmit the damage they sustain to other molecules and eventually to the vital targets. The vital targets are those molecules which must remain intact if the cell is to survive.

All this is quite straightforward, and much effort has been devoted to the identification of the vital and initial target molecules. There is now little doubt that the DNA is the most important vital target molecule [reviewed in 13]. It must, therefore, be assumed that cell death will result from radiation damage sustained by DNA directly, or indirectly from the actions of chemical species derived from the initial targets. This distinction between direct and indirect effects of radiation on DNA in cells has also been the subject of extensive investigations. Experiments with high concentrations of radical scavengers suggest that for sparsely ionizing radiations and in the presence of oxygen, the direct effect can contribute roughly between 30 and 40 percent of the total DNA damage [10]. However, these conclusions must be regarded with caution, because the use of molar concentrations of radical scavengers used to arrive at these estimates greatly modify the cellular environment, and because even these conditions are unlikely to lead to complete scavenging or radicals generated in close proximity to target molecules. This is particularly true for the important site-specific radical generation involving H_2O_2 and target-bound metal ions [11]. It is, therefore, likely that the contribution of direct radiation damage to DNA has been generally overestimated.

Since the probability of the interaction of ionizing radiation with target molecules in chemically complex systems is largely determined by their relative masses, in a typical cell about 70 percent of the high energy photons will interact with water. Water is not itself a vital target, but its decomposition by radiation is the most important route of indirect damage to other molecules in living cells. Biologically, the most important products of this decomposition are the hydroxyl (HO·) and super oxide (O_2^-) radicals, with the former generally held to be the most significant source of further damage because of its ability to oxidize most organic molecules at diffusion-controlled rates. This ability of HO· to attack indiscriminately makes the identification of the next molecule able to transmit damage to the DNA difficult. Most studies in this area have concentrated on the oxidation of lipids, because *in vitro* this can be an efficient process, involving a chain reaction. In theory, the oxidation of lipids in biological membranes could lead to changes in permeability and loss of integrity of the irradiated cell. However, comparative studies of the effects of incorporation of radioactive isotopes in DNA and outer cell membranes showed that the latter are not sensitive targets [14] and some more recent studies support this conclusion [1].

The role of the remaining major candidate molecules with potential to transmit the radiation-induced damage to DNA, the proteins, has attracted much less attention. Proteins are known to sustain damage as the result of attack by free radicals derived from water, but for many years it was believed that this was of little consequence to the cell [8]. Early studies of the effects of ionizing radiations on proteins showed the formation of cross links, chain fragmentation, damage to amino acid residues, increased proteolytic susceptibility, and loss of biological activity [2]. These are effectively passive changes, with their outcomes confined to the damaged proteins, which would in due course be replaced. However, this view had to be re-examined when we reported that irradiated proteins could acquire chemically reactive oxidizing and reducing groups, identified as hydroperoxides and dihydroxyphenylalanine (DOPA), respectively [12]. Subsequent studies of the peroxidized proteins showed that they could inactivate some antioxidants and enzymes, decompose to a range of new free radicals, and form covalent links with DNA [5, 6]. Thus, if the formation of protein hydroperoxides could be demonstrated in biological systems exposed to ionizing radiations, their ability to act as important transmitters of damage would have to be considered seriously.

We selected two kinds of biological systems for testing whether radiation could induce protein peroxidation: human blood serum and cultured cells. Previous investigations of the closely related system, blood plasma, showed that exposure to a flux of peroxyl radicals led to peroxidation of the plasma lipids and loss of antioxidants. Kinetic measurements showed that lipid oxidation was completely inhibited untill all the ascorbate (vitamin C) was exhausted by the action of the radicals [3]. This was an important finding, especially when damage to biological membranes was considered, because it implied that ascorbate could provide protection to biological fluids subjected to oxidant stress. Radiation is one form of such stress. Measurement of damage to proteins was not included in these studies. When we exposed serum to radiation-generated HO˙ radicals or to peroxyl radicals, we also found that the oxidation of the lipids was inhibited, as long as the serum contained ascorbate. After its exhaustion by the irradiation, lipid peroxides formed at a significant rate.

In contrast to the lipids, the oxidation of proteins in serum proceeded immediately on commencement of irradiation. This was measured by the assay of carbonyl residues and by the xylenol orange test for hydroperoxides [4]. Evidently, none of the endogenous serum antioxidants could prevent the formation of these reactive residues [9]. An interesting aspect of this study is the sequence in which the different components of the serum were oxidized. Since the oxidizing agent in these experiments was the HO˙ radical derived from the radiolysis of water, the probability of its reactions with different molecules can be roughly calculated from the composition of serum [7]. Thus, it can be shown that about 93 percent of these radicals react with proteins, up to 7 percent with lipids, and only 0.006 percent with ascorbate. As ascorbate disappeared rapidly from the start of the irradiations, its direct oxidation by HO˙ can be excluded. We, therefore, believe that these kinetic observations can be explained by a simple reaction sequence:

Start of irradiation (ascorbate present):

$$PrH + HO˙ \rightarrow Pr˙ + H_2O \tag{1}$$

$$Pr˙ + O_2 \rightarrow PrOO˙ \tag{2}$$

$$PrOO˙ + AH \rightarrow PrOOH + A˙ \tag{3}$$

After oxidation of ascorbate:
Reactions (1) and (2), followed by:

$$PrOO˙ + LH \rightarrow PrOOH + L˙ \tag{4}$$

$$L˙ + O_2 \rightarrow LOO˙ \tag{5}$$

$$LOO˙ + LH \rightarrow LOOH + L˙ \tag{6}$$

In this scheme PrH is a native protein, AH the unoxidised ascorbate, LH a fatty acid, and the dots indicate the corresponding free radicals. Reactions (5) and (6) represent the well-known chain peroxidation of lipids. It needs to be emphasized, that relatively large radiation doses (up to 1000 Gy) were used in these experiments, because of the limitations of the sensitivity of the assays available for peroxide measurements. However, these results demonstrated clearly that radiolysis of human blood plasma resulted in the formation of protein hydroperoxides which was not inhibited by any of the antioxidants present.

More recently, these studies were extended to cultured cells. In collaboration with a group in New Zealand, we showed the production of protein hydroperoxides in U937 cells treated with chemically generated peroxyl radicals. As in the case of serum, no lipid peroxides were found by the time of formation of micro-molar concentrations of protein hydroperoxides. This work was extended to a different line of cells, mouse hybrid myeloma, which were oxidized by radiation-generated hydroxyl as well as peroxyl free radicals. Again, the cell proteins were peroxidized long before the lipids.

The demonstration that protein hydroperoxides can form in relatively high yields in biological systems exposed to free radicals, typical of those produced in exposure to high-energy photons, suggests that they constitute a significant product of therapeutic irradiations. In addition, their ability to initiate potentially damaging processes identifies protein peroxides as a possible link in the events linking the exposure of a living organism to radiation and pathological changes. From the practical point of view, it will be important to discover whether this is indeed the case, and whether the formation of protein hydroperoxides can be influenced by deliberate means. If it can, physicians will be able to apply new procedures to enhance or to inhibit the effects of radiation on patients.

References

1. Caraceni, P. (1977) *Free Radical Biol. Med.* 23: 339–344.
2. Davies, M.J. and R.T. Dean (1997) *Radical Mediated Protein Oxidation.* Oxford; Oxford University Press.
3. Frei, B. *et al.* (1988) *Proc. Nat. Acad. Sci. USA* 85: 9748–9752.
4. Gay, C *et al.* (2000) *Redox Report.* 5: 55–56.
5. Gebicki, J.M. (1997) *Redox Report* 3: 99–110.
6. Gebicki, S. and J.M. Gebicki (1999) *Biochem. J.* 338: 629–636.
7. Gieseg, S. *et al.* (2000) *Biochem. J.* 350: 215–218.
8. Halliwell, B. (1988) *Biochem. Pharmacol.* 37; 569–571.
9. Lentner, C. (ed) (1984) Geigy Scientific Tables: Vol 3. Basle: Ciba-Geigy.
10. Ref [1], table 5.3, p 101.
11. Samuni, A. *et al.* (1985) *Eur. J. Biochem.*
12. Simpson, J.A. *et al.* (1992) *Biochem. J.* 282: 621–624.
13. Von Sonntag, C. (1987) *The Chemical Basis of Radiation Biology.* London: Taylor and Francis.
14. Warters, R.L. *et al.* (1977) *Curr. Topics Radiat. Res.* 12: 389–407.

Radiobiology and Bio-Medical Research
Edited by K.P. Mishra
Copyright © 2004 Narosa Publishing House, New Delhi, India

9. Effects of Auger-electron-emitting Radionuclides on Human Cells

F.H.A. Schneeweiss[1], S. Zerhusen[1] and R.N. Sharan[2]

[1]Institute of Medicine, Research Centre Jülich, 52425 Jülich, Germany,

[2]Department of Biochemistry, NEHU, Umshing, Shillong 793 022, India

Introduction

Auger-electron-(AE)-emitting radionuclides have the potential to be used in adjuvant therapy to augment the clinical efficacy of the conventional radiotherapy of tumors [4]. The main reasons for this are the emission of the large number (average 21 per decay in the condensed phase) of low-energy electrons [1] (from a few eV up to some tens of keV) and their extremely short range (from a few nm up to 10–20 μm). Therefore, when AE-emitting radionuclides such as ^{125}I are present in cells of neoplastic tissue, they would significantly damage these cells and spare the cells of the surrounding normal tissue. However, the clinical superiority of ^{125}I-based radiotherapy would depend on how efficiently the radionuclide is transported to the critical targets of the cells, for example, DNA [6]. Several ^{125}I-bound carriers are currently under examination to evaluate their biological efficiency.

The transport of AE-emitting radionuclides is also critical for proper dosimetry. For dosimetric calculation, ^{125}I nuclides would create problems if they were exclusively incorporated into DNA and the doses should be applicable for the cellular nucleus or the entire cell. The definition of radiation dose (energy deposition per mass of target) necessitates the energy being homogeneously distributed in the target. If ^{125}I is incorporated into DNA one can only determine the number of decays per target valid for the nuclide during a certain time range. Therefore, a suitable carrier molecule must be used which uniformly distributes ^{125}I inside the nucleus or the whole cell.

This investigation was designed to evaluate three different carriers for ^{125}I-induced cell damaging efficiency and dosimetric calculations. The experiments were carried out with asynchronous cultures of human kidney T1 cells at 37°C (physiological conditions). The following ^{125}I-labelled carrier molecules were used:

- ^{125}Iododeoxyuridine (^{125}IUdR), which would be incorporated into the DNA of S-phase cells,
- ^{125}Iodoantipyrine (^{125}I-AP) which would be distributed uniformly inside the cell, and
- Na ^{125}I, which would remain extracellularly located.

The damage induced by ^{125}I was measured by two biological end points: survival assay (clonogenic survival) and Comet assay (single cell electrophoresis) [2]. These ^{125}I-induced effects were compared with that found after ^{137}Cs-γ-irradiation.

Materials and Methods

Incorporation of [125] IUdR
Exponentially growing monolayer cultures of kidney T1 cells were incubated at 37°C and 5 percent CO_2 for 30 hours (approx. one population doubling time) with various ^{125}I concentrations of [125]IUdR (0, 74–15 kBq/ml medium). During this period of time the medium was renewed twice to prevent saturation of [125]IUdR uptake which was found to start between 8 and 10 hours. Fluorodeoxyuridine (FUdR) was added (10^{-8} M) to the medium to promote the incorporation of [125]IUdR into DNA by inhibition of the cellular thymidilate synthetase. FUdR increased [125]IUdR incorporation by a factor of ten. After the incubation period a single cell suspension was prepared and aliquots were measured in a γ-counter to calculate the ^{125}I decays per cell.

Exposure to ^{125}I-AP
Cells (5×10^6) were suspended in 2 ml medium of various ^{125}I concentrations of ^{125}I-AP (1.8-9.3 MBq), seeded into Leighton tubes and incubated for 28 hours at 37°C. It was assumed that ^{125}I-AP had uniformly distributed immediately after addition [8]. Calculations of decays per cell were based on the volume of T1 cells ($V = 3051$ μm^3, $\varnothing = 18$ μm).

Exposure to Na[125] I and[137]Cs-γ-rays
Corresponding to the [125]IUdR experiments T1 monolayers were exposed to various ^{125}I concentrations of Na[125]I (1.9–15 kBq/ml medium) without any further additives. The growth medium was also renewed twice during the period of ^{125}I exposure. At the end of this time no radioactivity was measured in the cells. For reference, 1×10^6 cells in suspension were irradiated on ice with ^{137}Cs-γ-rays (0.88 Gy/min) accumulating doses of up to 10 Gy.

Survival and Comet Assays
The cellular damage of the entire cell population was measured by survival assay reflecting the clonogenic ability of the cells. Eleven days after seeding, the colonies were stained with haematoxilin and counted (minimum of 50 cells). The alkaline Comet assay detected molecular damage mainly as DNA single-strand breaks. The assay was performed by a modified method of Singh [7]. Briefly, suspended cells in low-melting-point agarose were plated on agarose-precovered slides and lysed for 70 minutes at 4°C (lysing solution: 2.5 M NaCl, 100 mM Na_2-EDTA, 10 mM tris, one percent Na-sarcosinate, one percent Triton-X, pH 10). Subsequently DNA was allowed to unwind for 25 minutes in electrophoresis buffer (0.3 M NaOH, 1 mM Na_2-EDTA, pH >12) followed by a horizontal electrophoresis (0.8 V/cm) for 25 minutes. The negatively charged DNA fragments moved towards the anode and generated the so-called Comet

tail. After neutralization with 0.4 M tris-HCl (pH 7.5) the fragments were stained with propidium iodide (20 μg/ml) and analyzed by a fluorescence microscope connected to a PC hosting software (Komet 3.1, OPTILAS). The Olive Tail Moment (OTM) was chosen as the parameter to quantify the damage. It represents the product of percentage of DNA in the tail and the distance between the mean of the head and the mean of the tail [5]. The higher the DNA damage the more DNA fragments leave the head and migrate into the tail. This enhances the OTM value.

Results and Discussion

Effects of ^{125}IUdR and ^{125}I-AP

The survival curves (Fig. 1a) showed a steep decrease of the surviving fraction after ^{125}IUdR incorporation and a shoulder after ^{125}I-AP exposure. The D_{37} values calculated from the graphs were 90 d/c for ^{125}IUdR and 1160 d/c for ^{125}I-AP. The survival results after ^{125}IUdR incorporation were comparable with those found by Sedelnikova $et\ al.$ [6]. A characteristic feature of the ^{125}IUdR survival curve was the 'tailing' in the second decade of the logarithmic scale starting from approx. 1000 d/c [3]. The reason for this was a certain fraction of the asynchronous cell culture which did not pass the S-phase during incubation and was, therefore, responsible for preventing a further decline of surviving fractions. The consequence of this was an inhomogeneous distribution of ^{125}IUdR among the cells of the entire population. Theoretically, this resistant fraction could possibly be eliminated by an additional application of ^{125}I-AP which is independent of cell cycle specific problems of ^{125}IUdR incorporation. The steep linearly increasing functions of the Comet assay depicted stronger damage to DNA after ^{125}IUdR incorporation than after ^{125}I-AP exposure (Fig. 1b). Unlike the results obtained with the survival assay no 'tailing' was recognised by using the Comet assay after ^{125}IUdR incorporation (Figs. 1a & 1b). The non-labelled cell fraction

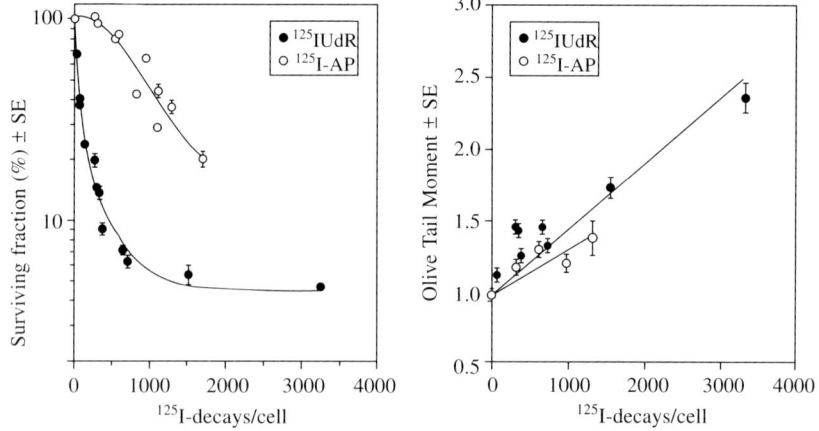

Figure 1. (a) Survival after ^{125}IUdR and ^{125}I-AP exposure ($n = 3$); (b) Comet assay after ^{125}IUdR-and ^{125}I-AP exposure ($n = 200$)

did not influence the Comet results. Whereas the survival assay determined the changed reproducibility of the entire cell population relative to the unaffected population, the Comet assay quantified the DNA damage of each individual cell as an absolute OTM value. This distinction is important After exposure of cells to ^{125}I-AP an acceptable correlation ($r = 0.76$) was found between the decreasing survival data and those of increasing DNA damage with the Comet assays (Figs. 1a & b).

Effect of Na^{125}I

After exposure to the highest radioactive concentration of Na^{125}I the surviving fraction only decreased to 90 percent of that of the controls (Fig. 2a). Also no significant differences of the OTM values compared with those of the controls could be analyzed with increasing Na ^{125}I concentrations (Fig. 2b). The slight decrease of the survival curve is probably caused by the 35.4 keV gamma rays which are emitted at a rate of 7 percent per ^{125}I decay of the extracellulary deposited Na ^{125}I. These photons and some released conversion electrons of higher energy can affect the intracellularly located DNA. This clearly demonstrates that the short-distant low energy AEs (93 percent up to 3.5 keV $\triangleq$ 400 nm) will reach the DNA if the AE-emitting nuclide is positioned close enough to the DNA depending on the carrier molecule. In the case of Na^{125}I the DNA target was hardly reached.

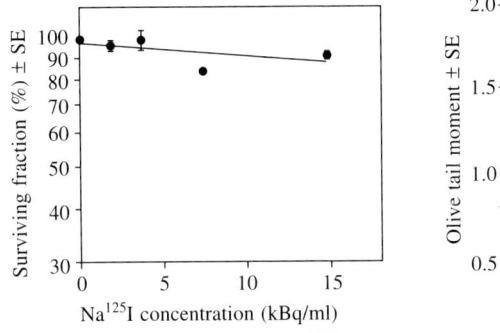 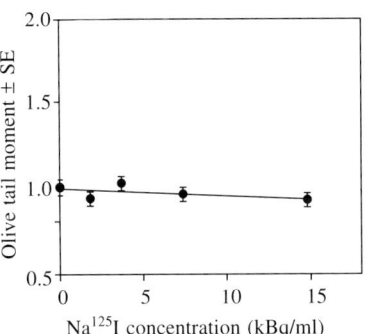

Figure 2. (a) Survival after Na ^{125}I exposure ($n = 3$); (b) Comet assay after Na ^{125}I exposure ($n = 200$)

Effect of ^{137}Cs-γ-irradation

Parallel experiments were carried out with ^{137}Cs-γ-irradiation in order to compare the ^{125}I induced results with those after exposure to γ-rays. The data obtained with the survival assay showed a decreasing curve down to the fourth decade of the logarithmic scale with a small shoulder and a D_0 of 0.9 Gy (Fig. 3a). The Comet assay results revealed an OTM-function increasing strictly linearly with the dose (Fig. 3b). Moreover, a correlation between the two assays (Fig. 3a and 3b) resulted in a high correlation coefficient of $r = 0.99$. Therefore it is postulated that after γ-irradiation the clonogenic survival of cells depends strongly on DNA/chromatin damage, particularly DNA strand breaks.

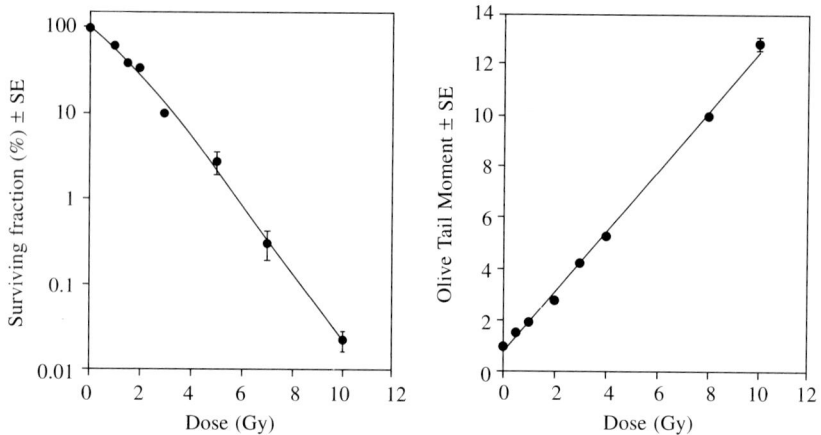

Figure 3. (a) Survival after ^{137}Cs-γ-irradiation ($n = 3$); (b) Comet assay after ^{137}Cs-γ-irradiation ($n = 200$)

Conclusion

Among the three different carrier molecules, IUdR, I-AP and NaI, each labeled with ^{125}I, ^{125}IUdR was found to be the most efficient in relation to DNA/chromatin damage and to the clonogenic survival of T1 and 86HG39 cells. The reason was the direct incorporation of ^{125}IUdR into the DNA whereas ^{125}I-AP distributed in the whole cell and Na^{125}I located only extracellularly were not bound to the target and were, therefore, less efficient in inducing destructive effects in the DNA and the cell. These results indicate that the selection of carrier is extremely important, especially for medical purposes. In the case of tumour therapy, ^{125}IUdR seems to be suitable as an adjuvant radiotherapeutic agent because the relative amount of S-phase cells of the fast proliferating tumour tissue is higher than in normal tissue and the strong cytotoxic effect of ^{125}IUdR can be generated even with small amounts of radioactivity. However, clinical use requires exact knowledge of the injected ^{125}I dose. An essential prerequisite for dosimetric calculation is the homogeneous energy deposition by uniform distribution of the radiopharmaceutical in the target. This condition is not fulfilled by ^{125}IUdR, but ^{125}I-AP is uniformly distributed over the entire cell including its nucleus. If the same number of decays per nucleus/cell is determined for both ^{125}I-AP and ^{125}IUdR and the correct dose (Gy) is calculated for the homogeneously distributed ^{125}I-AP then a comparable dose of ^{125}IUdR is obtained.

References

1. Charlton, D.E. and J. Booz (1981) A Monte Carlo Treatment of the Decay of ^{125}I. *Radiat. Res.* 87: 10–23.
2. Fairbairn, D.W., P.L. Olive and K.L. O'Neill (1995) The comet assay: a comprehensive review. *Mutation Res.* 339: 37–59.
3. Hofer, K.G., S.P. Bao (1995) Low-LET and high-LET radiation action of ^{125}I

decays in DNA: Effect of cysteamine on micronucleus formation and cell death. *Radiat. Res.* 141: 183–192.

4. O'Donoghue, J.A. and T.E. Wheldon (1996) Targeted radiotherapy using Auger Electron emitters. *Phys. Med. Biol.* 41: 1973–1992.

5. Olive, P.L., J.P. Banáth, R.E. Durand (1990) Heterogeneity in radiation-induced DNA damage and repair in tumor and normal cells measured using the comet assay. *Radiat. Res.* 122: 86–94.

6. Sedelnikova, O.A., I.G. Panyutin, A.R. Thierry and R.D. Neumann (1998) Radiotoxicity of Iodine-125-labeled oligodeoxyribonucleotides in mammalian cells. *J. Nucl. Med.* 39(8): 1412–1418.

7. Singh, N.P. (1988) A simple technique for quantitation of low levels of DNA damage in individual cells. *Exp. Cell Res.* 175: 184–191.

8. Talso, P.J., T.N. Lahr, N. Spafford, G. Ferenzi and H.R.O. Jackson (1955) A comparison of the volume of distribution of antipyrine, N-acetyl-4-amino-antipyrine, and ^{131}I-labeled 4-iodo-antipyrine in human beings. *J. Lab. Clin. Med.* 46: 619–623.

Radiobiology and Bio-Medical Research
Edited by K.P. Mishra
Copyright © 2004 Narosa Publishing House, New Delhi, India

10. Radiation Response of Oncogenes Associated with Signal Transduction

R.K. Bhattacharya

Department of Environmental Carcinogenesis and Toxicology
Chittaranjan National Cancer Institute, Calcutta 700 026, India

Abstract   Among a variety of biological effects in mammalian cells altered signal transduction has been recognized as a key cellular response to ionizing radiation. Several oncogenes, the products of which are components of signal transduction pathway and which are overexpressed in many tumors, are specifically induced in cells exposed to radiation. It has also become evident that the oncogene *ras* and the serine/threonine protein kinase oncogenes *raf* and PKC confer radioresistance to tumor cells. Modulation of these genes or their activity by natural compounds may offer strategy to treat cancer by enhancing radiation-induced apoptosis of tumor cells.

Introduction

The biological consequences of ionizing radiation include cell cycle arrest, chromosome aberration, gene mutation, cellular transformation and cell death [1–8]. Ionizing radiation also induces a transmissible chromosomal instability characterized by a variety of cytogenetic aberrations [9–11]. Most of these effects are attributed to the DNA damaging propensity of the irradiation resulting in irreversible changes during replication and/or repair of DNA [75, 53, 104]. The damage to DNA is also mediated through interaction of radiation with cell membrane [15, 17]. In recent years significant other responses to radiation have been recognized. Induction of specific cascades of genes is a key feature in cells exposed to radiation. Of particular importance are the oncogenes which play significant role in the development and progression of transformed cells. Certain oncogenes participate in cellular signal transduction pathways and are functionally regulated by signaling elements. Some of these genes also have been implicated in the cellular resistance to killing by ionizing radiation.

Oncogenes and Signal Transduction

Normal counterparts of many oncogenes (*proto-oncogenes*) are involved in vital normal cellular functions [7, 43]. They also interact with one another as components of signal transduction pathway, which involves transmission of messages from the membrane to the nucleus directing the cells to divide or to differentiate. Oncogene products include growth factors and growth factor receptors, various families of membrane associated tyrosine protein kinases, cytoplasmic serine/ threonine protein kinases, G-proteins and nuclear DNA binding proteins [13]. The family of *ras* oncogenes (H-, K- and N-*ras*) encodes for small membrane bound proteins of a molecular mass of 21 kDa that bind guanine nucleotides [9]. These proteins (*ras* p21) function as molecular switches that are turned on and off via a regulated GDP/GTP cycle [9, 25]. The *ras* p 21 protein has been implicated in the activation of phospholipase C [20] for generation of diacylglycerol (DAG), a second messenger that activates protein kinase C (PKC), a serine/ threonine protein kinase. Protein phosphorylation is a crucial event in the regulation of differentiation and proliferation [42], and PKC is a key element in signal transduction that controls cell growth and differentiation [81, 80]. The *ras* p21 also transduces signal through a more direct regulation of PKC [59]. PKC which is believed to be associated with tumorigenesis [79, 107] is a mediator for the products of several oncogenes, growth factors and cytokines as well as a number of tumor promoters, in these signaling pathways [8]. Besides PKC, the other cytoplasmic serine/threonine protein kinase genes *mos*, *cot* and *raf*-1 all have been associated with signal transduction. Of these oncogenes *raf*-1 which is downstream of *ras* in signal transduction pathway, in particular, appears to be at a central location in signal transduction pathways [77, 89]. The mitogenic signal received by *raf*-1 or other serine/threonine protein kinases including *ras* and mitogen activated protein (MAP) kinases is forwarded to the nucleus by phosphorylation of transcription factors such as AP-1, which is an association of the products of proto-oncogenes *fos* and *jun* [94] that regulate cell growth and proliferation.

Radiation Induces Protooncogene Expression

Exposure of low dose gamma radiation has been shown to induce the proto-oncogene c-*fos* in Syrian hamster embryo [111], Epstein-Barr virus transformed human lymphoblastoid 244B [32], NIH 3T3 [85] and primary rat tracheal epithelial [58] cells. Oncogenic transformation is a common feature in C3H 10T1/2 cells after gamma rays or He4 ions [82]. Transcriptional activation of c-*jun* has been observed in mammalian cells exposed to X-rays [65] and gamma rays [58, 65]. The activity of proto-oncogenes c-*myc*, c-H-*ras* [34] and c-*raf* [108] and that of tumor suppressor gene p53 [41] in several human cancer cell lines was upregulated by gamma ray exposure. In human breast cancer cell line MCF 7 the expression of several genes was elevated following irradiation with isosurvival doses of Fe ions or fission neutrons [3]. In animal tumors induced by ionizing radiation the oncogenes *ras* [98] and *myc* [27, 64] was shown to be amplified. Following whole body exposure of rodents to gamma rays expression of oncogenes c-*src*, c-H-*ras* [1], c-*fos* and c-*jun* [106] has been observed in various tissues. Overexpression of c-*fos* and c-*jun* also has been reported in the lymphocytes of mice after exposure to X-rays [62].

Radiation Activates Signal Transduction Events

Early responses of mammalian cells to ionizing radiation include the activation of PKC implicated in the regulation of gene expression, the stimulation of tyrosine protein kinase activity, and the enhancement of phosphatidylinositol (PI) turnover. Activation of PKC indeed is an early response to radiation exposure [8]. Low LET radiation (X-rays and gamma rays) induces dose-dependent increase in the expression of PKC mRNA in Syrian hamster embryo cells, while, high LET radiation (neutron) shows marginal effect only at low dose [112]. PKC activation occurs also in human tumor cells (leukemia and lung cancer) following exposure to gamma rays [35, 50]. The activation of PKC is functionally related to gene induction following exposure of mammalian cells to ionizing radiation [114]. Somewhat analogous to mitogen induced signaling, radiation stimulates tyrosine protein kinases and transcription factors [47] that modify *raf-1*, thus implicating *raf-1* in the ionizing radiation signal transduction pathway. Gamma irradiation also activates an Lck dependent signaling process through an *src* family tyrosine protein kinase in immature thymocytes [113]. Ionizing radiation, like tumor necrosis factor alpha (TNFα), induces rapid sphingomyelin hydrolysis to ceramide, stimulating ceramide-activated serine/threonine protein kinase in bovine aortic endothelial cells [34]. Activation of phospholipase D (PLD) transphosphatidylation has been observed indicating participation of PLD activation in signaling pathways in response to gamma radiation in human squamous carcinoma cells [2]. Information on signal transduction following whole body irradiation is scanty. It has been reported, however, that whole body irradiation with low dose X-rays increases intracellular calcium ions ($[Ca^{2+}]i$) and stimulates PKC activity in mouse lymphocytes [62, 63]. These findings suggested that low dose exposure could stimulate immunologic functions through facilitation of signal transduction process. In a recent study [54] We observed for the first time significant alterations in the PI signal transduction pathway in the liver of mice exposed to low dose of gamma rays. These alterations were reflected in the elevation of DAG level and concomitant increase in the activation of PKC. These early effects are of significance since these may set the stage for radiation-induced tumorigenesis and hence may be used to manipulate tumor response to radiotherapy. The signaling cascade in the cell has been known to respond also to other stimuli such as mitogens and chemicals. Chemical carcinogens, which display similar biological response as ionizing radiation, have been shown to alter PI turnover [18, 72], PKC activity modifying phosphorylation of key nuclear enzymes [73] and the activity of GTPase activating protein (GAP) leading to an overexpression of *ras* oncogene [19].

Radioresistance Associated with Overexpressed Signaling Elements

Certain tumor cell types have intrinsic resistance to killing by ionizing radiation, thus limiting radiation as a mode of cancer treatment. There exists a correlation between radioresistance of human tumors and the expression of specific oncogenes, namely the *ras* oncogene and the serine/threonine protein kinase oncogenes such as *raf, mos* and PKC. Transfection of NIH 3T3 cells with mutated N-, H- or K-*ras* or overexpressed H-*ras* was able to increase the resistance level of the recipient cell line [29, 83]. A synergistic increase in the radiation resistance level of primary rat embryo cells also was observed after cotransfection with *ras* and *myc* oncogenes [60, 68]. Normal human cells transfected with H-*ras* [99] and mutated *ras*-transformed human cells [71] as well as human [11] and rodent [37] tumor cells display radioresistance. The association was extended to include other oncogenes when transfection of NIH 3T3 cells by DNA from radiation resistant cells derived from a human laryngeal squamous carcinoma [46] and from radiation resistant noncancrous skin fibroblast cell line from cancer prone individuals [84, 14] led to the identification of an activated human *raf*-1 oncogene in the resultant radiation resistant transformants. The oncogene *raf*-1 was found to impart radioresistance also in human squamous carcinoma cells [48]. Transfections not only of the *raf*-1 oncogene but also of other serine/threonine protein kinase encoding oncogenes, *mos* and *cot*, have been shown to confer the radioresistant phenotype on the recipient NIH 3T3 cells [84] and hamster SHOK cells [100]. It has been proposed [110] that PKC plays a central role in a signal transduction "loop" affecting cellular response to radiation. Oncogene transformed radioresistant cells also display higher level of DAG [91], which is an endogenous activator of PKC. Other evidence [83] clearly indicated that the presence of the PKC-β1 cDNA was able to impart to NIH 3T3 cells an increased resistance to killing by gamma radiation. Moreover, specific oncogenes upstream (*sis, met, trk, ras*) and downstream (*ets* and *myc*) of the signaling mediators *raf, mos* and PKC also can induce the radiation resistance level of the cells [83].

Apoptosis and Radioresistance

Ionizing radiation kills cells by apoptosis [77–80]. Signaling pathways are associated with radiation-induced apoptosis. Early evidence suggested involvement of sphingomyelin signal transduction pathway for ionizing radiation-induced apoptosis. Ceramide generated after hydrolysis of membrane bound sphingomyelin stimulates a cascade of kinases and transcription factors that activate a common pathway of apoptosis [52]. Transfected c-*myc* and c-H-*ras* [16] and c-*fos* [86] also modulate radiation-induced apoptosis. Apoptotic cell death is severely restricted in radioresistant tumors and tumor cell lines [70]. Overexpression of protooncogene *bcl*-2, an anti-apoptotic gene, delays radiation-induced apoptosis in human cancer cells [56]. The lung cancer cell line U-1810 does not undergo apoptosis by radiation and expresses high *bcl*-2/*bax* ratio compared to radiosensitive cell lines U-1285 and U-1906 [96]. These studies also indicated that radiation-induced apoptosis was a p53 independent process and that overexpression of c-*myc* enhanced apoptosis in sensitive cells. Other results also confirm that overexpression of *bcl*-2 [67] or similar overexpression with low *myc* expression [89] correlated with resistance to radiation. The protooncogene *bax*, which forms a heterodimer with *bcl*-2 and accelerates apoptosis, is only weakly expressed in most breast cancer cells imparting radioresistance [92]. The tumor suppressor gene p53 is related to induction of apoptosis and suppression of wild type p53 expression has been shown to enhance the radiation resistance of human diploid fibroblasts. On the other hand, there was no evidence that loss of wild type p53 function increased the resistance of human tumor cells to ionizing radiation. Resistance to apoptosis plays an important role in tumors that are refractory to treatment by ionizing radiation. Modulating genes or their activity which impart radioresistance and/or use of heavy ion radiation [103] appear to have the advantage on apoptosis in radioresistant tumors.

Modulation of Signaling Elements Enhances Radiation-induced Apoptosis

It is apparent from the accumulated evidence that the oncogene *ras* and the serine/threonine protein kinase oncogenes such as *raf* and PKC confer radioresistance to tumor cells. Quantitative change in *ras* oncogene is commonly found in a wide variety of human neoplasms [10, 4]. This gene is activated through mutation [102] and the active oncoprotein *ras* p21 exists as GTP bound form [9, 25] which is negatively regulated by GAP through stimulation of hydrolysis of GTP [25]. On the basis of evidence obtained from carcinogen-induced *ras* activation we proposed that GAP phosphorylated by activated PKC helps keep the *ras* p21 in the more active GTP bound state [19]. The cysteine residue at the fourth position from the COOH terminal of *ras* proteins undergoes enzymatic farnesylation [30], the recognition sequence for farnesyltransferase in H-*ras* p21 being CVLS [76]. Isoprenylation (with farnesyl or geranylgeranyl group) enables *ras* proteins to attach themselves to plasma membrane and to exert their oncogenic activity. Oncogenic *ras* proteins lose their transforming activity when such attachment is blocked by mutation in the COOH terminus recognition sequence [12]. Pharmacological blockade of prenylation has been used to inhibit *ras* action. Inhibition of oncogenic *ras* activity by specific inhibitors of prenylation has been shown to reduce the radiation survival of a variety of human tumor cells [6]. The farnesyltransferase inhibitor also has been shown to radiosensitize H-*ras* transformed rat embryo fibroblasts [5]. It has been discussed earlier that PKC is a key element in signal transduction [81, 80], associated with tumorigenesis [79, 107], and regulates *ras* p21 signal transduction [59]. The activity of this enzyme is enhanced in tumors [101, 102 and is activated by translocation from cytosol to membrane. PKC also plays a role in *ras* activation [19]. Hence modulation of PKC is expected to improve radiation resistance of cells. Inhibition of PKC indeed has been shown to sensitize human tumor cells to ionizing radiation [35, 50]. Chelerythrine, a PKC α and β inhibitor [36] enhances ionizing radiation-induced apoptosis in radioresistant tumor cells. This is mediated by increasing sphingomyelinase activity to produce more ceramide [15]. In addition, calphostin, another PKC inhibitor [51], also enhances radiation-induced apoptosis in a similar manner [21, 22]. Inhibition of PKC and subsequent ceramide production alters the expression and function of antiapoptotic proteins [21] and activates caspases (CED 3/CPP 32 proteases), which are required for ionizing radiation-mediated apoptosis [23]. These studies suggest that inhibition of PKC may represent a strategy to enhance ionizing radiation-mediated cell death by apoptotic mechanism in cells that are otherwise resistant to radiation.

Conclusion and Future Direction

There is a considerable void in the information with regard to the status on signal transduction and oncogene expression in response to heavy ion irradiation. Since high LET radiation has important application in radiotherapy specific strategies can be evolved with a view to sensitize cells. The primary objective is to minimize radiation dose for effective induction of apoptotic cell death. Agents designed to inhibit the activation of ras or PKC could achieve this objective, since these are the primary signal transduction elements that impart radioresistance. Overexpression of both these genes is also evident in tumor cells. Inhibition of ras activity in tumors is difficult to achieve, and since activation of ras can be mediated through phosphorylation of GAP [19], modulation of PKC would be the most appropriate strategy to increase the apoptotic potential of cell. Most PKC inhibitors are toxic to normal cells. It is important to screen natural compounds which can selectively inhibit PKC activation. Nicotinamide, an endobiotic, has recently been shown to confer radiosensitivity [40, 39]. It also has been found to prevent overexpression of ras as well as activation of PKC in animals following carcinogen administration [20]. Curcumin, the yellow pigment of the commonly used spice turmeric, has been the subject of extensive research over the past decade. It is toxicologically very safe and exhibits many interesting biological and pharmacological properties and is a prospective cancer chemopreventive agent [28]. Recently, curcumin has been observed to induce apoptosis in human leukemic cell lines [90]. Rutin, a naturally occurring glycoside of the flavonol quercetin, has a protective role against carcinogenesis [109]. Both curcumin and rutin are effective inhibitors of PKC [73]. Lycopene, a carotenoid abundantly present in tomato, also has been found recently to inhibit activation of PKC [55]. Studies could be extended to screen other PKC modulators of natural origin and to evaluate their efficacy to enhance radiation-induced apoptosis in tumor cells. This could form a rational strategy for the treatment of resistant tumors.

Acknowledgement

My work reported herein was conducted during my tenure at Bhabha Atomic Research Centre and for which I am grateful to Dr. Malini Krishna and several students. I sincerely thank Prof. (Dr.). M. Siddiqi for his interest.

References

1. Anderson, A. and G.E. Woloschak (1992) *Radiat. Res.* 130: 340–344.
2. Avila, M.A., G. Otero, J. Cansado, A. Dritschilo, J.L. Velasco and V. Notario (1993) *Cancer Res.* 53: 4474–4476.
3. Balcer-Kubiczek, E.K., X.F. Zhang, G.H. Harrison, X.J. Zhou, R.M. Vigneulle, R. Ove., W.A. McCready and J.F. Xu (1999) *Int. J. Radiat. Biol.* 75: 529–541.
4. Barbacid, M. (1987) *Annu. Rev. Biochem.* 56: 779–927.
5. Bernhard, E.J., G. Kao, A.D. Cox, S.M. Sebti, A.D. Hamilton, R.J. Muschel and W.G. McKenna (1996) *Cancer Res.* 56: 1727–1730.
6. Bernhard, E.J., W.G. McKenna, A.D. Hamilton, S.M. Sebti, Y. Qian, J.M. Wu and R.J. Muschel (1998) *Cancer Res.* 58: 1754–1761.

7. Bishop, J.M. (1991) *Cell* 64: 235–248.

8. Blumberg, P.M. (1991) *Mol. Carcinog* 4: 339–344.

9. Bollag, G. and F. McCormic (1991) *Annu. Rev. Cell. Biol.* 7: 601–633.

10. Bos, J.L. (1988) *Mutat Res.* 195: 255–271.

11. Bruyneel, E.A., G.A. Storme, D.C. Scallier, D.L. Van den Berge, P. Hilgard and M.M. Mareel (1993) *Eur. J. Cancer* 29A: 1958–1963.

12. Buss, J.E., P.A. Solski, J.P. Schaeffer, M.J. MacDonald and C.J. Der (1989) *Science* 243: 1600–1602.

13. Cantley, L.C., K.R. Auger, C. Carpenter, B. Duckworth, A. Graniani, R. Kepeller and S. Soltof (1991) *Cell* 64: 281–302.

14. Chang, E.H., K.F. Pirollo, Z.Q. Zou, H.-Y. Cheung, E.L. Lawler, R. Garner, E. White, W.B. Bernstein, Jr. J.W. Fraumeni and W.A. Blatner, (1987) *Science* 237: 1036–1039.

15. Chaumra, S.J., H.J. Mauceri, S. Advani, R. Heimann, M.A. Beckett, E. Nodzenski, J. Quintans, D.W. Kufe and R.R. Weichselbaum (1997) *Cancer Res.* 57: 4340–4347.

16. Chen, C.H., J. Zhang and C.C. Ling (1994) *Radiat. Res.* 139: 307–315.

17. Choudhary, D., M., Srivastava, A. Sarma and R.K. Kale, (1998) *Radiat. Env. Biophys.* 37: 177–185.

18. Choudhury, S., M. Krishna and R.K. Bhattacharya (1995) *Cancer Lett* 93: 213–218.

19. Choudhury, S., M. Krishna and R.K. Bhattacharya (1996) *Cancer Lett* 109: 149–154.

20. Choudhury, S., M. Krishna and R.K. Bhattacharya (1995) *Cancer Lett.* 147: 39–44.

21. Chumra, S., E. Nodzenski, M. Crane, D. Hallahan, R. Weichselbaum and J. Quintans (1996) *Adv. Exptl. Med. Biol.* 406: 39–45.

22. Chumra, S.J., E. Nodzenski, J. Quintans and R.R. Weichselbaum (1996) *Cancer Res.* 56: 2711–2714.

23. Datta R., D. Banach, H. Kojima, R.V. Talanian, E.S. Alnemri, W.W. Wong and D.W. Kufe (1996) *Blood* 88: 1936–1943.

24. Domen, J., K.L. Gandy, and I.L. Weissman (1998) *Blood* 91: 2272–2282.

25. Downward, J. (1992) *Bioessays* 14: 177–184.

26. Elkind, L.M. and H. Sutton (1959) *Nature* 184: 1293–1295.

27. Felber, M., F.J. Burns and S.J. Garte (1992) *Radiat. Res.* 131: 297–301.

28. Firozi, P.F., V.S. Aboobaker and R.K. Bhattacharya (1995) *Chem-Biol. Interact.* 100: 41–51.

29. Fitz Gerald, T.J., L.A. Rothstein and C. Daugherty (1985) *Am. J. Clin. Oncol.* 8: 517–522.

30. Glomset, J.A. and C.C. Farnsworth (1994) *Annu. Rev. Cell Biol.* 10: 181–205.

31. Gorgojo, L. and J.B. Little (1989) *Int. J. Radiat. Biol.* 55: 619–630.

32. Gouldwell, W.T., J.H. Ulm and I.P. Antel (1991) *Neurosurgery* 29: 880–887.

33. Guillem, J.G., C.A. O'Brian and C.J. Fitzer (1987) *Cancer Res.* 47: 2036–2039.

34. Haimovitz-Freidman, A., C.C. Kan, D. Ehleiter, S.S. Persaud, M. McLoughlin, J. Fuks and R.W. Kolesnick (1994) *J. Exp. Med.* 180: 525–535.

35. Hallahan, D.E., S. Virudachalam, J.L. Schwartz, N. Panje, R. Mustafi and R.R. Weichselbaum (1992) *Radiat. Res.* 129: 345–350.

35. Harms-Ringdahl, M., P. Nicotera and I.R. Radford (1996) *Mutat. Res.* 366: 171–179.

36. Herbert, J.M., J.M. Augereau, J. Gleye and J.P. Maffrand (1990) *Biochem. Biophys. Res. Commun* 172: 993–999.

37. Hermens, A. and P. Bentvelzen (1992) *Cancer Res.* 52: 3073–3082.
38. Herskind, C., S. Hass, M. Flentje and E.W. Hahn (1996) *Radiat. Res.* 145: 299–303
39. Horsman, M.H. (1995) *Acta Oncol.* 45: 167–174.
40. Horsman, M.H., D.W. Sieman, D.J. Chaplin and J. Overgaard (1997) Radiother Oncol, 45: 167–174.
41. Huang, H., C.Y. Li, and J.B. Little (1996) *Int. J. Radiat. Biol.* 70: 151–160.
42. Hunter, T. and J.A. Cooper (1985) *Annu Rev. Biochem.* 54: 897–930.
43. Hunter, T. (1991) *Cell* 64: 249–270.
44. Kadhim, M.A., S.A. Lorimore, M.D. Hepburn, D.T. Goodhead, V.J. Buckle and E.G. Wright (1994) *Lancet* 344: 987–988.
45. Kadhim, M.A., S.A. Lorimore, K.M.S. Townsend, D.T. Goodhead, V.J. Buckle and E.G. Wright (1995) *Int. J. Radiat. Biol.* 67: 287–293.
46. Kasid, U., A. Pfeifer, R.R. Weichselbaum, A. Dritschilo and G.E. Mark (1987) *Science* 237: 1039–1041.
47. Kasid, U., S. Suy, P. Dent, S. Ray, T.L. Whiteside and T.W. Sturgill (1996) *Nature* 382: 813–816.
48. Kasid, U., A. Pfeifer, T. Brennan, M. Breckett, R.R. Weichselbaum, A. Dritschilo and G.E. Mark (1989) *Science* 243: 1354–1356.
49. Katain, M. and P.J. Parker (1988) *Nature* 232: 203.
50. Kim, C.Y., A.J. Giaccia, B. Strulovici and J.M. Brown (1992) *Br. J. Cancer* 66: 844–849.
51. Kobayashi, E., K. Ando, H. Nakano, T. Iida, H. Ohno, M. Morimoto and T. Tamaoki (1989) *J. Antibiotic* 42: 1470–1474.
52. Kolesnick, R.N., A. Haimovitz-Friedman and Z. Fuks (1994) *Biochem. Cell Biol.* 72: 471–474.
53. Kraft, G. (1987) *Nucl Sci. Appl.* 3: 1–28.
54. Krishna, M., K. Pasupathy, J.G. Satav and R.K. Bhattacharya (1999) *Ind. J. Exptl. Biol.* 37: 1075–1079.
55. Krishna, M. (2000) Unpublished observation.
56. Kyprianou, N., E.D. King and D. Bradbury (1997) *Int. J. Cancer* 70: 341–348.
58. Lagroye, I. and J.L. Poncy (1998) *Bioelectromag* 19: 112–116.
59. Lacal, J.C., P.T. Fleming, B.S. Warren, P.M. Blumberg and S.A. Aaronson (1987) *Mol. Cell Biol.* 7: 4146–4149.
59. Leyko, W. and G. Bartosz (1986) *Int. J. Radiat. Biol.* 49: 743–770.
60. Ling, C.C. and B. Endlich (1989) *Radiat. Res.* 120: 267–279.
61. Little, J.B., L. Gorgojo and H. Vetros (1990) *Int. J. Radiat. Oncol. Biol. Phys.* 19: 1425–1429.
62. Liu, S.J., X. Su. Y.C. Zhang and Y. Zhao (1994) *Chin Med. J.* 107: 431–436.
63. Liu, S.Z., X. Su, Z.B. Han, Y.C. Zhang and J. Qi (1994) *Biomed. Environ. Sci.* 7: 284–291.
64. Lumniczky, K., S. Antal, E. Unger, E.J. Hidvegi and G. Safrany (1997) *Radiat. Oncol Invest* 5: 158–162.
65. Manome, Y., R. Datta, N. Taneja, T. Shafinan, E. Bump, R. Hass, R. Weichselbaum and D. Kufe (1993) *Biochemistry* 32: 10607–10613.
67. McCarthy, N.J., S.A. Hazelwood, D.S. Huen, A.B. Rickinson and G.T. William (1996) *Adv. Exptl. Med. Biol.* 406: 83–97.
68. McKenna, W.G., M.C. Weiss, B. Endlich, C.C. Ling, V.J. Bakanauskas, M.L. Kelsten and R.J. Muschel (1990) *Cancer Res.* 50: 97–102.
69. Meyn, R.E. (1997) *Oncology* 11: 319–359.
70. Meyn, R.E., L.C. Stephens and L. Milas (1996) *Cancer Meta. Rev.* 15: 119–131.

71. Miller, A.C., K. Kariko, C.E. Myers, E.P. Clark and D. Samid (1993) *Int. J. Cancer* 53: 302–307.
72. Mistry, K.J., M. Krishna and R.K. Bhattacharya (1996) *Chem-Biol. Interact* 98: 145–152.
73. Mistry, K.J., M. Krishna, K. Pasupathy, V. Murthy and R.K. Bhattacharya (1996) *Chem. Biol. Interact* 100, 177–185.
73. Mistry, K.J., M. Krishna and R.K. Bhattacharya (1997) *Cancer Lett.* 121: 99–104.
75. Mookherjee, A. and S.B. Bhattacharjee (1984) *Aspects of Radiation Biophysics.* New Delhi: Interprint.
76. Moores, S.L., M.D. Schaber, S.D. Mosser, E. Rands, M.B. O'Hara, V.M. Garsky, M.S. Marshall, D.L. Pompliano and J.B. Gibbs (1991) *J. Biol. Chem.* 266: 14603–14610.
77. Morrison, D.K. (1990) *Cancer Cells* 2: 377–380.
78. Mothersill, C., K. O'Malley, J. Harney, F. Lyng, D.M. Murphy and C.B. Seymour (1997) *Radiat. Res.* 147: 156–165.
79. Nishizuka, Y. (1984) *Nature* 308: 693–697.
80. Nishizuka, Y. (1989) *J. Am. Med. Assoc.* 262: 1826–1833.
81. Nishizuka, Y. (1992) *Science* 253: 607–614.
82. Piao, C.Q. and T.K. Hei (1993) *Carcinogenesis* 14: 497–501.
83. Pirollo, K.F., Y.A. Tong, Z. Villegas, Y. Chen and E.H. Chang (1993) *Radiat Res.* 135: 234–243.
84. Pirollo, K.F., R. Garner, S.Y. Yang, L. Li, W.A. Blattner and E.H. Chang (1989) *Int. J. Radiat. Biol.* 55: 783–796.
85. Prasad, A.V., M. Mohan, B. Chandrasekar and M.L. Meltz (1995) *Radiat. Res.* 143: 263–272.
86. Puck, T.T. and P.I. Marcus (1956) *J. Exptl. Med.,* 103: 653–666.
86. Pruschy, M., Y.Q. Shi and N.E. Crompton (1996) *Biochem. Biophys. Res. Commun.* 241: 519–524.
88. Raju, M.R. (1980) Heavy Particle Radiotherapy (Academic Press, New York).
89. Rapp, U.R. (1991) *Oncogene* 6: 495–500.
90. Roy, M. and R.K. Bhattacharya (2000) Unpublished observation.
91. Ruggiero, M., F. Casamassima, L. Magnelli, S. Pacini, J.H. Pierce, J.S. Greenberger and V.P. Chiarugi (1992) *Biochem. Biophys. Res. Commun,* 183: 652–658.
92. Sakakura, C., E.A. Sweeny, T. Shirahama, Y. Igarashi, S. Hakamori, H. Nakatani, H. Tsujimoto, T. Imanashi, M. Ohgaki, T. Ohyama, J. Yamazaki, A. Hagiwara, K. Sawai and T. Takahashi (1996) *Int. J. Cancer* 67: 101–105.
93. Samid, D., A.C. Miller, D. Rimoldi, J. Garner and E.P. Clark (1991) *Radiat. Res.* 126: 244–250.
94. Sassone-Corsi, P., W.W. Lamph, M. Kamps and I.M. Verma (1988) *Cells* 54: 553–560.
95. Seymour, C.B., C. Mothersill and T. Alper (1986) *Int. J. Radiat. Biol.* 50: 167–179.
96. Sirzen, F., B. Zhivotovsky, A. Nilsson, J.H. Bergh and R. Lewensohn (1998) *Anticancer Res.* 18: 695–699.
97. Sklar, M.D. (1988) *Science* 239: 645–647.
98. Sloan, S.R. and A. Pellicer (1992) *Prog. Clin. Biol. Res.* 374: 1–18.
99. Su, L.-N. and J.B. Little (1992) *Int. J. Radiat. Biol.* 62: 201–210.
100. Suzuki, K., M. Watanabe and J. Miyoshi, (1992) *Radiat. Res.* 129: 157–162.
101. Szumiel, I. (1994) *Int. J. Radiat. Biol.* 66: 329–341.
102. Tabin, C.J., S.M. Bradley, C.L. Bargmann, R.A. Weinberg, A.G. Papageorge,

E.M. Scolnick, R. Dhar, D.R. Lowy and E.A. Chang (1982) *Nature* 300: 143–149.

103. Takahashi, T., N. Mitsuhashi, M. Furuta, M. Hasegawa, T. Ohno, Y. Saito, H. Sakurai, T. Nakano and H. Nibe (1998) *Anticancer Res.* 18: 253–265.

104. Taucher-Scholz, G., J. Heilmann, and G. Kraft (1996) *Nucl. Instr. Meth. Phys. Res. B.* 107: 318–322.

105. Thompson, L.H. and H.D. Suit (1969) *Int. J. Radiat. Biol.* 15: 347–362.

106. Usenius, T., M. Tenhunen and J. Koistinaho (1996) *Neuroreport* 7: 2559–2563.

107. Vertosick, Jr. F.T. (1992) *Cancer J.* 5: 328–331.

108. Warenius, H.M., M. Jones, M.D. Jones, P.G. Browning, L.A. Seabra and C.C. Thompson (1998) *Br. J. Cancer* 77: 1220–1228.

109. Webster, R.P., M.D. Gawde and R.K. Bhattacharya (1996) *Cancer Lett.* 109: 185–191.

110. Weichselbaum, R.R., D.E. Hallahan, V. Sukhatme, A. Dritschilo, M.L. Sherman and D.W. Kufe (1991) *J. Natl. Cancer Inst.* 83: 480–484.

111. Woloschak, G.E., C.M. Chang-Liu, P. Shearin-Jones and C.A. Jones (1990) *Cancer Res.* 50: 339–348.

112. Woloschak G.E., C.-M. Chang-Liu and P. Shearin-Jones (1990) *Cancer Res.* 50: 3963–3967.

113. Wu, G., J.S. Danska and C.J. Guidos (1996) *Int. Immunol* 8: 1159–1164.

114. Young, C.Y., P.E. Murtha and J. Zhang (1994) *Oncol Res.* 6: 203–210.

Radiobiology and Bio-Medical Research
Edited by K.P. Mishra
Copyright © 2004 Narosa Publishing House, New Delhi, India

11. Very High Level Natural Radiations Areas (VHLNRAs) and Current Radiation Protection Policies: Controversies, Beliefs and Facts

S.M.J. Mortazavi[1,2*]

[1] National Radiation Protection Department (NRPD), Iranian Nuclear Regulatory Authority (INRA), P.O. Box 14155-4494, Tehran, Iran

[2] Medical Physics Department, School of Medicine, Rafsanjan University of Medical Sciences (RUMS), Rafsanjan, Iran

Abstract: People in some areas around the world live in dwellings with radiation and radon levels as much as 100 times the global average. Inhabited areas with high levels of natural radiation are found in different areas around the world including Yangjiang, China; Kerala, India; Guarapari, Brazil and Ramsar, Iran. Ramsar in northern Iran is among the world's well-known areas with highest levels of natural radiation. Annual exposure levels in areas with elevated levels of natural radiation in Ramsar are up to 260 mGy y^{-1} and average exposure rates are about 10 mGy y^{-1} for a population of about 2000 residents. Due to the local geology, which includes high levels of radium in rocks, soils, and groundwater, Ramsar residents are also exposed to high levels of alpha activity in the form of ingested radium and radium decay progeny as well as very high radon levels (over 1000 MBq m^{-3}) in their dwellings. In some cases, the inhabitants of these areas receive doses much higher than the current ICRP-60 dose limit of 20 mSv y^{-1}. As the biological effects of low doses of radiation are not fully understood, the current radiation protection recommendations are based on the predictions of an assumption on the linear, no-threshold (LNT) relationship between radiation dose and the carcinogenic effects. Considering LNT, areas having such levels of natural radiation must be evacuated or at least require immediate remedial actions. Interestingly, most local physicians in Ramsar report anecdotally there is no increase in the incidence rates of cancer or leukemia in their area. Inhabitants of the high level natural radiation areas (HLNRAs) of Ramsar are largely unaware of natural radiation, radon, or its possible health effects, and the inhabitants have not encountered any harmful effects due to living in their paternal houses. In this regard, it is often difficult to ask the inhabitants of HLNRAs of Ramsar to carry out remedial actions. Using LNT and ALARA, public health in HLNRAs like Ramsar is best served by relocating the inhabitants. However, the residents' health seems unaffected and relocation is upsetting to the residents. It can be concluded that in HLNRAs the LNT model might be inappropriate to use as the basis for public-health measures.

Introduction

Humans, animals and plants have been exposed to natural radiation since the creation of life. The annual per capita effective doses from natural and Man-made sources for the world's population is currently about 2.8 mSv. Nearly 85 percent of this dose (2.4 mSv) comes from natural background radiation (UNSCEAR 2000). Levels of natural radiation can vary greatly. Ramsar (Fig.1), a northern coastal city in Iran, has some areas with one of the highest levels of natural radiation studied so far. The effective dose equivalents in very high level natural radiation areas (VHLNRAs) of Ramsar in particular in Talesh Mahalleh, are few times higher than the dose limits for radiation workers. Inhabitants who live in some houses in this area receive annual doses as high as 132 mSv from external terrestrial sources.

Figure 1. **Ramsar, a city in northern Iran, has some inhabited areas with highest levels of natural radiation in the world.**

External exposure rates from terrestrial gamma radiation in Iran and the annual background doses to the inhabitants of some areas around the world are summarized in Tables 1 and 2 respectively.

Table 1. **External exposure rates from terrestrial gamma radiation in Iran**

Iran's Important Radiological Data	
Population in 1996 (10^6)	69.98
Average absorbed dose rate in air (nGy h^{-1}): Outdoors	71
Average Absorbed dose rate in air (nGy h^{-1}): Indoors	115
Indoors/outdoors ratio	1.6

Source: Survey of natural radiation exposure, UNSCEAR 2000.

Radioactivity of the high level natural radiation areas (HLNRAs) of Ramsar is due to ^{226}Ra and its decay products, which have been brought up to earth

Table 2. Mean and maximum annual natural terrestrial radiation doses to the inhabitants of some areas around the world.

Country	Area	Approximate population	Absorbed Dose rate in air[a] (nGy h^{-1})
Brazil	Guarapari	73 000	90–170 (street)
			90–90 000 (beaches)
Iran	Ramsar[b]	2 000	70–17 000
India	Kerala	100 000	200–4 000
China	Yangjiang	80 000	370 (average)

(a) includes cosmic and terrestrial radiation.
(b) it should be noted that the monazite sand beaches at Guarapari in Brazil have a higher dose rate, but these areas are uninhabited. Therefore it can be claimed that Ramsar has the highest level of natural radioactivity studied so far.
(Source: UNSCEAR 2000).

surface by the water of warm springs. There are more than 9 hot springs with different concentrations of radium in this city. The visitors as well as residents usually use these springs as spas. According to the results of the surveys performed by the Atomic Energy Organization of Iran (AEOI), the radioactivity seems to be first due to the mineral water and second due to some travertine deposits that have thorium content higher than that of uranium (14). As the biological effects of low doses of radiation are not fully understood, the current radiation protection recommendations are based on the predictions of an assumption on the linear, no-threshold (LNT) relationship between radiation dose and the carcinogenic effects. Considering the LNT theory as a scientific fact, there is a general belief that even low levels of radiation as well as exposures to natural sources are harmful. In spite of the fact that at present time we have no considerable radio-epidemiological data regarding the incidence of cancer in the inhabitants of VHLNRAs of Ramsar, some of the local physicians strongly believe that the population living in these areas does not reveal increased solid cancer or leukemia incidence. As the majority of the inhabitants of Ramsar live there since many generations, we started a study to assess whether there is an apparent lack of radiation susceptibility among residents of the high level natural radiation areas.

Chromosome Aberrations

Cytogenetic analysis of chromosome aberrations in peripheral blood lymphocytes is widely used to detect and quantify exposure to radiation and other clastogens. Previous cytogenetic studies have shown a statistically significant difference between the results of the inhabitants of HLNRAs and a normal background radiation area (NBRA) [4]. However, our results showed no significant difference even in the case of the inhabitants who lived in houses with extraordinary elevated levels of natural radiation. Although at present there are no substantial epidemiological data on the incidence of cancer among the inhabitants, local physicians strongly believe that the persons who live in the HLNRAs of Ramsar do not show any increased cancer, or leukemia. As this might be attributed to the difference in individual susceptibilities to radiation exposure, we conducted an experiment to assess the possible existence of a radioadaptive response in the inhabitants of these areas.

Dose-Effect Relationship

There is a great controversy about the dose-effect relationship in published reports on the frequency of chromosome aberrations induced by chronic exposure to elevated environmental levels of radiation. This controversy exists in studies of residents in areas with elevated levels of natural radiation as well as the residents of areas contaminated by nuclear accidents. Using chromosomal aberrations as the main endpoint, we carried out an experiment to assess the dose-effect relationship in the residents of high level natural radiation areas of Ramsar. A cytogenetical study was performed on 21 healthy inhabitants of the high level natural radiation areas and 14 residents of a nearby control area. Our preliminary results showed no positive correlation between the frequency of chromosome aberrations and the cumulative dose of the inhabitants.

Adaptive Response

Epidemiological evidence has indicated that the natural radiation in HLNRAs is not harmful to residents. Furthermore, cancer mortality rate is significantly lower in the high background areas than in the control areas [5]. This is one of typical examples of radiation hormesis. To find out whether a high level natural radiation acts as an adapting dose, we may study the radio-resistance of blood lymphocytes of residents in these areas after irradiation of samples with high doses [5]. In a preliminary experiment we found that when the lymphocytes of the inhabitants of HLNRAs and a neighboring NBRA are exposed to 1.5 Gy gamma rays, the frequency of chromosome aberrations in the lymphocytes of the inhabitants of HLNRAs is significantly ($P < 0.001$) lower than that of the neighboring NBRA [11].

Hematological Alterations

Since 1999, National Radiation Protection Department (NRPD) of the Iranian Nuclear Regulatory Authority (INRA) has studied the effects of prolonged high-level natural irradiation on hematological and immunological parameters in man. In this regard, healthy donors of both sexes who lived in VHLNRAs as well as donors from a neighboring normal background radiation area (NBRA) were examined for hematological changes. Hematological parameters such as counts of leukocytes (WBC), lymphocytes, monocytes, granulocytes, red blood cells (RBC), hemoglobin (Hb), hematocrit (Ht), MCV, MCH, MCHC, RDW, PLT, and MPV were studied in all of the individuals. Our results indicated that there is no any statistically significant alteration in hematological parameters of the inhabitants of VHLNRAs of Ramsar and the neighboring control area.

Immunological Changes

It is a well known that high doses of ionizing radiation suppress the activity of the immune system. On the other hand, the low-level whole body irradiation (WBI) can enhance the immunological response. To assess whether relatively high doses of natural radiation can alter humoral immune parameters, an experiment was conducted on the inhabitants of VHLNRAs of Ramsar, permanently living in houses with elevated levels of natural radiation. Immunological factors such as the concentration of serum immunoglobulins of IgA, IgG, IgM and C3, C4 components of the complement system in healthy donors from VHLNRAs and a neighboring NBRA were studied. Our findings indicate that there is a slight increase in IgA and IgG levels of the inhabitants of VHLNRAs compared to those of matched controls. IgM, C3, and C4 complements were in the normal range. In spite of the fact that the increase in IgA and IgG were not so marked to show probable enhanced immunological capability, it can be concluded that relatively high doses of natural radiation are not immuno-suppressive. More research is needed to clarify the immunological alterations induced by different levels of natural radiation.

HLNRAs and the Current Controversies

Substantial evidence indicate that it is possibly incorrect to estimate the hazard of the low radiation doses and very low dose rates by straight extrapolation of the effects of much higher doses and dose rates higher by more than 10 orders of magnitude, such as encountered by the survivors of nuclear attacks in Hiroshima and Nagasaki. Radio-epidemiological studies on the inhabitants of HLNRAs provide a unique opportunity to study effects of relatively high doses at low dose rates, such as experienced in the normal practice of radiological protection. Due to statistical considerations, these studies should rather be of long duration. In Ramsar, the population that lives in the HLNRAs is estimated to be about 2,000 persons. In this regard, to obtain statistically reliable results, only a long-term study can provide considerable number of person-years of observation. On the other hand there are published reports on the increased life-span of A-bomb survivors [9] or the increased survival of laboratory animals exposed to low doses of ionizing radiation [7]. Therefore, the life-span of the inhabitants should be studied as a part of the future long-term studies.

The risk of low-dose radiation exposures has for a variety of reasons been highly politicized. This has led to a frequently exaggerated perception of the potential health effects, and to lasting public controversies [8]. Current radiation protection recommendations are based on the predictions of an assumption on linear, no-threshold dose-effect relationship (LNT). Beneficial effects and lack of detriment after irradiation with low levels of ionizing radiation, including a prolonged exposure to high levels of natural radiation of the inhabitants of HLNRAs, are inconsistent with LNT. In HLNRAs of Yangjiang county in China (annual doses are about 330 mR) it has been indicated that mortality from all cancers and those from leukemia, breast and lung were not higher than that of the control area (110 mR/y). Furthermore, it was shown that when samples of circulating lymphocytes taken from the inhabitants were tested *in vitro* for mitotic response to phytohemagglutinin (PHA) and the degree of unscheduled DNA synthesis (UDS), there were higher responsiveness and UDS rates in the HLNRAsamples than in those from the control area [1]. It was found that in a HBRA in China the cancer (non-leukemia) mortality was 14.6 percent lower than in NBRA, and the leukemia mortality among men was 15 percent lower and among women 60 percent lower. No difference in the frequency of various genetical diseases was observed between Chinese HBRA and NBRA [17]. These hormetic effects contradict the linear no-threshold theory. Similar health surveys should be performed in the HLNRAs of Ramsar, where the annual doses are much higher than the Yangjiang County. There are many other areas with high levels of background radiation around the world, and epidemiological studies have indicated that natural radiation in these areas is not harmful for the inhabitants. It can be reconfirmed that a threshold separates the health effects of natural radiation from the harm of large doses. This threshold seems to be much higher than the greatest level of natural radiation (lifetime doses up to 19.6 Sv in VHLNRAs of Ramsar).

Conclusion

Using LNT and ALARA, public health is best served by relocating HLNRAs inhabitants. Several statistically significant epidemiological studies contradict the validity of LNT concept by showing hormetic effects in a form of risk decrements of cancer mortality and mortality from all causes in populations exposed to low-dose radiation [12, 11]. Populations in areas with high level natural radiation show no adverse health effects when compared to low-dose populations. Furthermore, several studies of large populations indicate beneficial health effects of low doses of ionizing radiation, lower mortality and disease rates [2, 3, 18, 6]. Our preliminary findings on the biological effects of prolonged exposure to high levels of natural radiation in the inhabitants of VHLNRAs of Ramsar, showed no harmful health effects. It can be concluded that in HLNRAs the LNT model might be inappropriate to use as the basis for public health measures.

Acknowledgement

The author would like to thank Dr. K.P. Mishra, whose comments led to a much improved paper. I would also like to acknowledge Dr. M. Ghiassi-Nejad, the director general of Iranian NRPD for his interest, and encouragement. He is also thankful to R. Assaie, A. Heidary, R. Varzegar, F. Zakeri, A. Esmaili, Dr. Asghari and Dr. M. Jafari and their colleagues for valuable participation in this study.

References

1. Chen, D. and L. Wei (1991) Chromosome aberration, cancer mortality and hormetic phenomena among inhabitants in areas of high background radiation in China. *J. Radiat. Res.* (Tokyo), 32 Suppl 2: 46–53.
2. Cohen, B.L. (1995) Test of the linear no-threshold theory of radiation carcinogenesis for inhaled radon decay products. *Health Phys.* 68: 157–174.
3. Cohen, B.L. (1996) Problems in the radon versus lung cancer test of the linear no-threshold theory and a procedure for resolving them. *Health Phys.* 72: 623–628.
4. Fazeli, T.Z. (1990) Cytogenetic studies of inhabitants of a high level natural radiation area of Ramsar. Proceeding of International Conference on High Levels of Natural Radiation (ICHLNR), Ramsar., Iran, 459–464.
5. Ikushima, T. (1999) Radioadaptive response: responses to the five questions. *Human and Experimental Toxicology* 18: 433–435.
6. Jagger, J. (1998) Natural background radiation and cancer death in Rocky Mountain states and Gulf Coast states. *Health Phys.* 75: 428–430.
7. Jaworowski, Z. (1997) Beneficial effects of radiation and regulatory policy. *Australas Phys. Eng. Sci. Med.* 20(3): 125–38.
8. Kellerer, A.M. (2000) Risk estimates for radiation-induced cancer—the epidemiological evidence. *Radiat. Environ. Biophys.* 39(1): 17–24.
9. Mine, M., Y. Okumura, M. Ichimaru, T. Nakamura, and S. Kondo (1990) Apparently beneficial effect of low to intermediate doses of A-bomb radiation on human lifespan. *Int. J. Radiat. Biol.* 58(6):1035–43.
10. Mortazavi, S.M.J., T. Ikuhima, H. Mozdarani and A.A. Sharafi (1999) Radiation Hormesis and Adaptive Responses Induced by Low Doses of Ionizing Radiation. Journal of Kerman University of Medical Sciences 6, (1): 50–60.
11. Mortazavi, S.M.J., M. Ghiassi Nejad and M. Beitollahi (2001) Very High Background Radiation Areas (VHBRAs) of Ramsar: Do We Need any Regulations to Protect the Inhabitants? Proceedings of the 34th midyear meeting, Radiation Safety and ALARA Considerations for the 21st Century, California, USA, 177–182.
12. Pollycove, M. (1998) Nonlinearity of radiation health effects. *Environ Health Perspect*, 106 Suppl. 1(9806): 363–8.
13. Roth, J., P. Schweizer, and C. Guckel (1996) Basis of radiation protection. *Schweiz Med. Wochenschr.* 126 (26): 1157–71.
14. Sohrabi, M. (1990) Recent radiological studies of high level natural radiation areas of Ramsar. Proceeding of International Conference on High Levels of Natural Radiation (ICHLNR), Ramsar, Iran, 3–7.
15. Sohrabi M. and A. Esmaili (2000) Recent radiological studies of high level natural radiation areas of Ramsar. Fifth International Conference on High Levels of Natural Radiation (ICHLNR), Munchen, Germany.
16. United Nations Scientific Committee on the Effects of Atomic Radiation (2000) Sources and Effects of Ionizing Radiation, UN, New York.
17. Wei, L., Y. Zha, Z. Tao, D. Chen and Y. Yua (1990) Epidemiological investigation and radiological effects in high background radiation areas of Yangjiang, China. *Radiation Research* 31: 119.
18. Wei, L. (1997) High background radiation area-an important source of exploring the health effects of low dose ionizing radiation. In: *High Levels of Natural Radiation: Radiation Dose and Health Effects.* Wei, L.; Sugahara, T.; Tao, Z., ed. *Beijing, China; Amsterdam.* The Netherlands: Elsevier: 1–66.

Radiobiology and Bio-Medical Research
Edited by K.P. Mishra

12. Human Milk Caseins: A Review

Satish M. Sood and Charles W. Slattery

Department of Biochemistry, Loma Linda University, School of Medicine,
Loma Linda, CA 92350, USA

Abstract: Human milk proteins, and particularly the caseins, play not only a nutritional role in the newborn but may also produce a number of other important biological effects. The elucidation of protein structure may help in the understanding of these functions. Caseins are complex in nature and cannot be crystallized for X-ray crystallography. Also, the determination of casein structure by NMR spectroscopy is at or beyond the current limits of the technique. α-Casein is present in human milk at only about 5 percent of the casein fraction and thus has been studied very little. The β-casein of human milk, its major casein constituent (~75 percent), is unique in that it may be phosphorylated to any level from 0 to 5 organic phosphates per molecule. The forms with 2 and 4 phosphates are the major components with about 30 percent each. All six forms of human β-casein have been isolated and purified. They all have the same amino acid sequence of 211 amino acid residues and a molecular mass of about 25 kDa. The phosphorylation sites are also known. Detailed study of these six different forms reveal some similarities and some interesting differences. Sedimentation and viscosity data yielded a solvation of 2.9 g H_2O/g protein and an axial ratio of about 5 for the monomer. This would be consistent with a prolate ellipsoid of 10 nm length and 2 nm width. Increasing the temperature from 4 to 37°C causes conformational changes and an increase in protein aggregation which is further enhanced by the addition of NaCl at this temperature *until* a limiting size is reached. This polymer contains 95–105 monomers and is nearly spherical with a radius of about 15 nm. κ-Casein is about 18 percent of the casein in human milk and is highly glycosylated (~55 percent). Galactose and N-acetyglucosamine constitute ~ 35 percent of the carbohydrate. With a molecular mass of 35.5 kDa, 158 amino acid residues and one organic phosphate per molecule, it stabilizes all the different forms of human β-casein from precipitation in the presence of 10 mM Ca^{2+}. Genetic engineering now holds out the hope that these proteins may soon be produced in large enough quantities not only to study adequately but to feed infants that can not obtain mothers milk for various reasons and provide them with the biological advantages of human milk proteins.

Introduction

Milk is secreted by the mammary gland and is the major source of various nutrients for the newborn. Human milk is recommended as the best food for the infant due to many unique properties that benefit the breast-fed infant. In all mammalian species so far examined, the major phosphoproteins of milk occur as colloidal aggregates referred to as casein micelles. These complexes also contain the majority of the inorganic calcium and phosphate of milk, which will otherwise precipitate and would not be delivered to the infant. The structure of human casein micelles is important from a nutritional point of view because it is related to their function of delivering proteins and minerals to the infant in a liquid suspension that is readily ingested and, furthermore, one that is easily digested to form components that may be efficiently absorbed into the body. Structure is also important from a technological point of view for the manufacturing of infant milk formulae.

Milk Proteins

The proteins of milk are divided into two fractions: [1] the caseins, which are suspended as complexes of proteins, carbohydrates and minerals, and [2] the soluble or whey proteins. Major casein proteins in milk are designated α, β, γ, and κ-caseins from their electrophoretic mobility while the whey proteins are mainly α-lactoalbumin and lactoferrin in human milk and β-lactoglobulin in other species. Early milk is rich in immunoglobulins from the mother. There are also some important enzymes [lysozyme, α-amylase, protease] and binding proteins [folate, vitamin B_{12}-binding] in human and other milks.

Casein Micelles

Two characteristics of human casein micelles that distinguish them from most other species are: [1] their average size, which is about one-third to one-half those of cow's milk (Fig. 1) as determined by electron microscopy [6, 8], and [2] their solvation, which is about twice that of cow's milk micelles [28]. Since there was still some question about the size because of measuring artifacts, diameters of nearly 200 human milk micelles and 400 bovine milk micelles were measured using scanning tunneling microscopy or STM (Fig. 2 [27]. Bovine casein micelles had an average diameter of 230 nm and those in the human milk averaged about

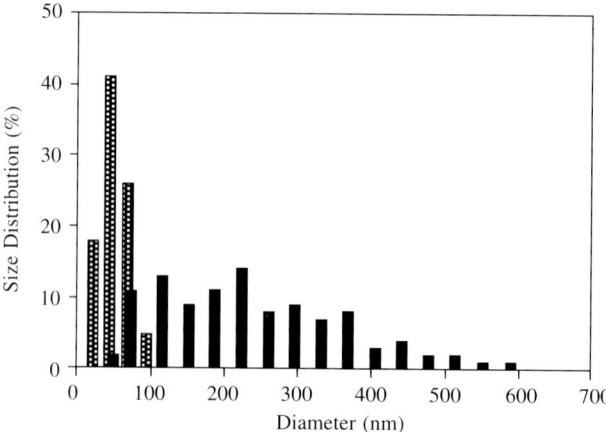

Figure 1.   Size distribution of bovine and human casein milk micelles.

50 nm. The association between the micelles seemed random Micelles were closely packed and the number of nearest neighbors were determined by only steric considerations. In other words, micelles did not appear to have 'sticky spots' on their surfaces. Previously published electron micrographs clearly indicated the existence of distinct submicellar structures not seen with STM. These preparations were only air-dried, in contrast to the thorough drying necessary for electron micrographs. It is assumed that each micelle is covered with a layer of carbohydrate-rich material from κ-casein which may be extensively hydrated. This probably lends a smooth appearance to the micelles.

Figure 2.   A cluster of human casein micelles of different sizes as seen by Scanning Tunneling Microscopy.

Casein Structure

α-Caseins

Caseins are typical proteins. Due to their hydrophobic nature, caseins tend to aggregate and cannot be crystallized. Although α_{s1}-casein is the major protein in cow's milk [~45 percent], and there were some earlier studies, including two-dimensional gel electrophoresis, indicating its presence in human milk [1, 34], the subject, however, debatable. Recently, its presence has been confirmed [7, 19], but only in minor amounts [~5 percent] with an apparent M_r of 27,000 [19]. Human α_{s1}-casein has been reported to exist naturally as a multimer in complex with κ-casein in mature human milk, thereby being unique among α_{s1}-caseins [19]. The presence of three cysteines in the protein provides a molecular explanation for the interactions in the complex [17]. This protein was purified by a combination of gel-filtration and ion exchange chromatography under denaturing conditions. Amino acid sequence analysis results showed a high level homology with bovine α_{s1}-casein and demonstrated that the protein contained at least two cysteine residues (19).

β-Caseins

Monomer Properties

Human β-casein, the major component of the casein system in human milk, has a molecular weight of ~25,000 Da [12], at or beyond the current limits for structure determination by NMR spectroscopy. Consequently, other techniques such as analytical ultracentrifugation, viscosity, laser light scattering, fluorescence spectroscopy and atomic force microscopy have been used to investigate its structure [22, 33]. Using a combination of sedimentation coefficient and reduced viscosity data for the monomeric human β-caseins, one is able to calculate a solvation value of 2.9 g H_2O/g protein and an axial ratio of about 5 [22]. This would be consistent with a prolate ellipsoid of 11 nm in length and 2 nm in width.

Calcium Binding

Phosphorylated residues in the caseins play an important role in maintaining the size of the micelles in milk [36]. Human β-casein is multi-phosphorylated, has six different forms with 0-5 phosphate groups/molecule [13]. The order of phosphorylation is at serine residues 10/9, 9/10, 8, and 6, and then the threonine residue at position 3 [12]. The doubly-phosphorylated [2-P] and quadruply-phosphorylated [4-P] forms are the most abundant, each being between 25 and 35 percent of the total (Fig. 3) [22]. Phosphorylation of β-casein is of biological importance in the assembly of stable micelles [36]. The charged phosphorylated residues of the casein molecules maintain micelle size, and play an important role in the formation of the soft curd of human milk [36]. They also have a subsequent nutritional function in that they allow the caseins to bind divalent metal ions [Ca^{2+}, Zn^{2+}, Mn^{2+}, Fe^{2+}, etc.] and any associated inorganic phosphate or other anions [4, 11, 14, 18]. They precipitate in the presence of Ca^{+2} in the absence of a stabilizing factor and are thus designated as calcium-sensitive proteins. Rapid dialysis equilibrium with a radioactive $^{45}CaCl_2$ label was used to determine the dissociation constant for binding of Ca^{+2} to the human β-caseins with different levels of phosphorylation [22–25, 30, 31]. It was concluded that there is one

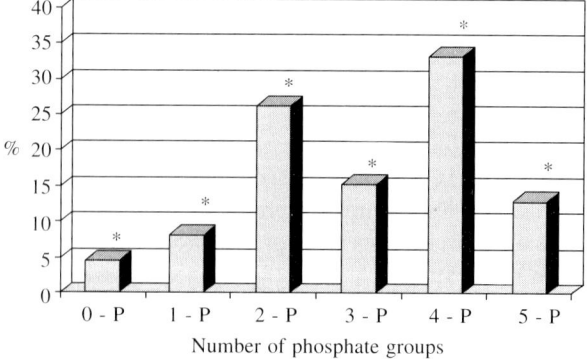

Figure 3. Percentage of each β-casein from nine different samples of human milk.

strong binding site for Ca^{+2} for each organic phosphate ester. The dissociation constants for the 1-P and 2-P forms are significantly higher than those obtained for bovine β-casein. This may be important for calcium absorption by the newborn human infant.

Aggregation Characteristics

The β-casein monomers because of the hydrophobic nature tend to be monomeric at 4°C and aggregate into polymers as the temperature is increased to near the physiologic value of 37°C [22]. Calcium ions aid the aggregation by binding to the negatively charged phosphates and reducing electrostatic repulsion [22]. As the monomers aggregate as the temperature is increased, the reduced viscosity (a measure of particle shape) decreases and the particles apparently become less ellipsoidal and more symmetrical [22]. There seems to be a transition temperature, above which the sedimentation coefficient (a measure of particle size) increases rapidly [22]. The addition of Ca^{+2} at even higher concentrations causes the transition temperature, where the phosphorylated human β-caseins begin major aggregation and precipitation, to occur at ever lower temperatures, as determined by turbidity measurements (33). The value for the sedimentation coefficient in the absence of Ca^{+2} at 37°C is inversely related to the degree of phosphorylation of the molecule, confirming the influence of electrostatic repulsion by the charged phosphates that are concentrated toward the N-terminal portion of the molecule. However, at 37°C, addition of NaCl to raise the ionic strength and reduce electrostatic interactions, mimics the effects of Ca^{+2} binding without introducing the possibility of cross-linking and results in a rapid decrease in the reduced viscosity to stable low values and a rapid increase in the sedimentation coefficient to stable high values which are the same for all levels of monomer phosphorylation [22, 25, 30, 31]. The values translate into a spherical polymer with a radius of about 15 nm and a solvation of approximately 3 g H_2O/g protein containing about 90 monomers. These values are confirmed by dynamic light scattering [16, 31]. This polymer may represent the minimum-sized micelle in the human system.

κ-Casein

Human κ-casein is much more highly glycosylated [~55 percent] [9] than bovine κ-casein [~5 percent] [21]. Another important difference between the human and bovine entities is that the human protein contains only one cysteine residue (5) while the bovine protein contains two [21]. Thus, bovine κ-casein may form disulfide-linked polymers while human κ-casein may not. It may only form dimers at a maximum. κ-Casein is a calcium-insensitive protein and its major role is to prevent β-casein precipitation in the presence of Ca^{+2} and to limit the growth of the micelle (2, 3, 35). Measurements of the ability of human κ-casein to prevent the precipitation of human β-caseins by 5 or 10 mM Ca^{+2} showed that as little as 0.15 moles of κ-casein for each mole of β-casein could completely stabilize the calcium-sensitive protein against precipitation [20]. The relationship of κ-casein content and micelle size for human milk micelles has been investigated. There is an inverse relationship between both κ-casein content and κ-casein glycosylation and micelle size (10).

A Possible Micelle Model

Monomer Structure

We have seen that the β-casein monomers may be approximated by a prolate ellipsoid about 11 nm long and 2 nm wide (Fig. 4). The N-terminus would contain the charged phosphothreonine and/or phosphoserines, along with other charged residues, while the C-terminus is largely hydrophobic resulting in an amphipathic molecule. Similar considerations could apply to the κ-casein monomers. In this case, it is the C-terminus of the molecule that is cleaved away by *rennin* to give the highly glycosylated glycomacropeptide and the N-terminus is hydrophobic. As with the bovine κ-casein [15], this would result in an amphipathic structure.

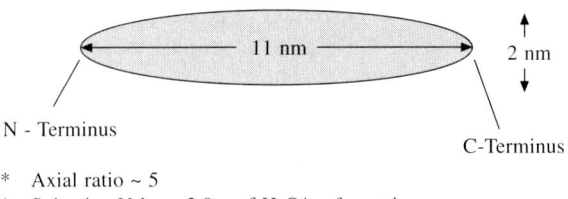

N - Terminus   C-Terminus

*   Axial ratio ~ 5
*   Solvation Value - 2.9 g of H_2O/g of protein

Figure 4. **Dimensions of the β-casein monomer as determined by hydrodynamic methods.**

Submicelle or Minimum Micelle Structure

We have also seen that β-casein molecules may aggregate to form spherical polymers containing about 80–90 monomers. In these, the hydrophobic portion of the molecule would be toward the interior and the hydrophilic charged N-terminus on the surface (Fig. 5). It seems reasonable, given the similar amphipathic

Figure 5.   **Schematic representation of the mode of association of human β-casein.**

nature, that κ-casein monomers could replace some β-casein monomers and also fit into this polymer with the hydrophobic N-terminus of the molecule inward and the glycosylated hydrophilic C-terminus on the surface. If there was a relatively large amount of κ-casein in a polymer and the distribution of κ-and β-caseins was fairly even, further aggregation of the polymer to participate in larger sized micelles would be prevented. This would constitute a minimum sized micelle. However, if there was only a small amount of κ-casein in a polymer and/or an uneven distribution of the κ- and β-caseins such that the κ-casein only constituted a small portion of the surface (Fig. 6), the polymer could participate in further aggregation to form larger micelles (Fig. 7) through hydrophobic interaction when charge is reduced by calcium binding or through calcium-bridging between organic phosphates on different polymers. These would thus constitute submicellar entities. The size to which the micelle could grow would be determined in an inverse relationship by the amount of κ-casein it contained.

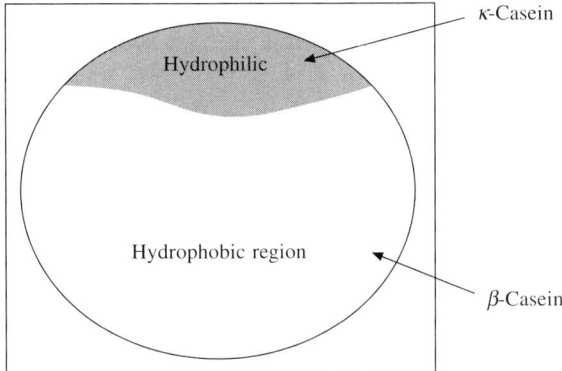

Figure 6. Schematic representation of a possible human casein submicelle containing both β- and κ-caseins with an uneven distribution. The hydrophobic region contains the β-casein and the hydrophilic region contains the κ-casein.

Figure 7. Diagram showing the growth of submicelles to a larger micellar size.

References

1   Anderson, N.D., M.T. Powers and S.L. Tollasksen (1982) Proteins of Human Milk. I. Identification of major components. *Clin. Chem.*, 28: 1045–1055.

2. Azuma, N., S. Kaminogawa and K. Yaumauchi (1984). Molecular weight and conformation of human κ-casein and its interactions with other human milk proteins. *Agric. Biol. Chem.* 48: 771–76.

3. Azuma, N., S. Kaminogawa and K. Yamauchi (1985) Reconstitution of human casein micelle and its properties. *Agric. Biol. Chem.* 49: 2655–60.

4. Blakeborough, P., D.N. Saler and M.I. Gurr. (1983) Zinc binding in cow's milk and human milk. *Biochem. J.* 209: 505–512.

5. Brignon, G., A. Chtourou and B. Ribadeau Dumas (1985) Preparation and amino acid sequence of human κ-casein. *FEBS lett.* 188: 48–54.

6. Calapaj, G.G. (1962) Ricerche al microscopio ellettonico sulla polidispersione della caseina di latte umano e bovine. *Boll. Sier. Oter milan* 41: 276–291.

7. Cavaletto, M.A., G. Contisani, L.N. Giuffride and A. Conti (1994) Human α_{s1}-casein like protein: purification and N-terminal sequence determination. *Biol. Chem. Hoppe-Seyler* 375: 149–151.

8. D'Agostini, B.A. and G.G. Calapaj (1958) Ricerche al m.e. sulla caseina del latte di alcune specie animali. *Acta. Med. Vet.* 4: 9–15.

9. Dev, B.C., S.M. Sood, DeWind, S. and C.W. Slattery (1993) Characterization of human κ-casein purified by FPLC. *Preparative Biochemistry.* 23: 389–407.

10. Dev, B.C., S.M. Sood, S. DeWind and C.W. Slattery (1994) κ-Casein and β-casein in human milk micelles: structural studies. *Arch. Biochem. Biophys.* 314: 329–336.

11. Dickson, I.R. and D.J. Perkins (1971) Studies on the interactions between purified bovine caseins and alkaline earth metal ions. *Biochem. J.* 124: 235–240.

12. Greenberg, R., M.L. Groves and H.J. Dower (1984) Human β-casein: amino acid sequence and identification of phosphorylation sites. *J. Biol. Chem.* 259: 5132–5138.

13. Groves, M.L. and W.G. Gordon (1970) The major component of human casein: a protein phosphorylated at different levels. *Arch. Biochem. Biophys.* 140: 47–51.

14. Harzer, G. and H. Bauer (1982) Binding of zinc to casein. *Am. J. clin. Nutr.* 35: 981–987.

15. Hill, R.J. and R.G. Wake (1969) Amphiphile nature of κ-casein as the basis for its micelle stabilizing property. *Nature.* 221: 636–639.

16. Javor, G.T., S.M. Sood, P. Chang and C.W. Slattery (1991) Interactions of triply phosphorylated human β-casein: fluorescence spectroscopy and light-scattering studies of conformation and self-association. *Arch. Biochem. Biophys.* 289: 39–46.

17. Johnson, L.B., L.K. Rasmussen, T.E. Peterson and L. Berglund (1995) Characterization of three types of human alpha s_1-casein mRNA transcripts. *Biochem. J.* 309: 237–42.

18. Lönnerdal, B. and L.S. Hurley (1985) Manganese binding proteins in human and cow's milk. *Am. J. Clin. Nutr.* 41: 550–559.

19. Rasmussen, L.K., H.A. Due and T.E. Petersen (1995) Human α_{s1}-casein: purification and characterization. *Comp. Biochem. Physiol.* 111B: 75–81.

20. Rollema, H.S. (1992) Casein association and micelle formation. In *Advances in Dairy Chemistry.* [Edit or P.F. Fox,] Vol. 1: pp 111–140. London: Elsevier Applied Science.

21. Ruettimann, K.W. and M.R. Ladisch (1987) Casein micelles: structure, properties and enzymatic coagulation. *Enzyme Microb. Technol* 9: 578–598.

22. Sood, S.M., P. Chang and C.W. Slattery (1985) Interactions in human casein systems: self association of fully phosphorylated human β-casein. *Arch. Biochem. Biophys.* 242: 355–364.

23. Sood, S.M., P. Chang and C.W. Slattery (1988) Interactions in human casein systems: self association of non-phosphorylated human β-casein. *Arch. Biochem. Biophys.* 264(2): 574–583.

24. Sood, S.M., P. Chang and C.W. Slattery (1990) Interactions of triply phosphorylated human β-casein: Monomer characterization and hydrodynamic studies of self association. *Arch. Biochem. Biophys.* 277: 415–421.

25. Sood, S.M., P. Chang and C.W. Slattery (1992a) Interactions properties of doubly phosphorylated β-casein: a major component of the human milk caseins. *J. Dairy Sci.* 75: 2937–2945.

26. Sood, S.M., P.J. Herbert and C.W. Slattery (1997) Structural studies on casein micelles of human milk: dissociation of β-casein of different phosphorylation levels induced by cooling and ethylenediaminetetraacetate. *J. Dairy Sci.* 80: 628–633.

27. Sood, S.M., G. Javor, K. Koller, V. Ellings and C.W. Slattery (1992b) Bovine and human casein micelles as seen by scanning tunneling/atomic force microscopy. *FASEB J.* 7: A158.

28. Sood, S.M., K.S. Sidhu and R.K. Dewan (1979) Heat stability and Voluminosity and hydration of casein micelles form milk of different species. *New Zealand J. Dairy Sci. and Tech.* 14: 217–225.

29. Sood, S.M. and C.W. Slattery (1989) Association of human β-caseins. Mixture of the two major components. *J. of Cell Biol.* 107: 202a.

30. Sood, S.M. and C.W. Slattery (1993) Interaction properties of singly phosphorylated human β-casein: similarities with other phosphorylation levels. *J. Dairy Sci.* 77: 405–412.

31. Sood, S.M. and C.W. Slattery (1997) Monomer characterization and studies of self-association of the major β-casein of human milk. *J. Dairy Sci.* 80: 1554–1560.

32. Sood, S.M. and C.W. Slattery (2000) Association of the quadruply-phosphorylated β-casein from human milk with the nonphosphorylated form. *J. of Dairy Sci.* (in press).

33. Slattery, C.W., S.M. Sood and P. Chang (1989) Hydrophobic interactions in human casein micelle formation; β-casein aggregation. *J. Dairy Res* 56: 427–33.

34. Voglino, G.F. and A. Ponzone (1972) Polymorphism in human casein. *Nature* 238: 149–150.

35. Yamauchi, K., N. Azuma, H. Kobayashi and S. Kaminogawa (1981) Isolation and properties of human κ-casein. *J. Biochem* 90: 1005–1012.

36. Yun, S.E., K. Ohmiya and S. Shimizu (1982) Role of the phosphoryl groups of β-casein in milk curdling. *Agric. Biol. Chem* 46: 1505–1511.

Radiobiology and Bio-Medical Research
Edited by K.P. Mishra
Copyright © 2004 Narosa Publishing House, New Delhi, India

13. Patient Dose Reduction and Quality Assurance Evaluation in Whole Body CT Scanner

S.P. Mishra, Vinita Gupta, P. Narayan and Shalini Jaiswal

Department of Medical Physics, Regional Cancer Centre,
Kamala Nehru Memorial Hospital, Allahabad-211002, India

Low level X-ray radiation inmating from CT diagnostic procedures have potential to produce deterministic effect. Patient organ dose data for CT examinations for various anatomical sites have been a matter of investigation over last few years to quantify the probable harm associated with it and ways to reduce it. The emphasis has been to reduce the organ dosage without compromising the spatial resolution of the CT Scanner and impending clinical efficacy. CT Scanners use high kV and mAs pencil X-ray beams which have ability to contribute a large dose to patients. The spatial resolution of CT Scanner is proportional to the X-ray dose and improvement in sensitivity could only be achieved at the expense of increased dose or decreased resolution. The uncertainty in CT pixel or noise is thus proportional to number of transmitted X-ray photons ($\Box$ n), that is, dose to organ of interest.

The evaluation of organ doses data have a special significance in pregnancy, paediatrics and tubercloma (where repeated CT scans are required) and all abdomino-pelvic CT procedures in child bearing age group. The data enables the probabilistic harm assessment with the procedure and hence charting out preventive measures. The quality assurance (QA) of the CT Scanner has a bearing on dose reduction and the investigating ability of the CT Scanner for clinical relevance. Thus there is a need to analyse the dosage required for a quality CT Scan image and its optimal performance for an acceptable CT parameters at reduced entrance surface dose.

It can be seen from aforesaid relation that for a given dose and spatial resolution there is a fixed sensitivity in CT scan image to tissue changes set by the physics of measurement. It, therefore, becomes essential to maintain the CT performance and quality assurance parameters to reduce the organ doses to patients. In this study methods to reduce the organ dosage have been discussed and evaluation of QA parameter for optimal performance identified. An all purpose comprehensive CT phantom has been developed to study these aspects of the spatial resolution and organ dosage.

The performance of CT system, image quality, radiation dose and radiation protection which contribute in one way or the other to the final picture are of intrinsic importance in quality image formation. Hence, prior to acceptance of the system, thorough evaluation of the system's performance as per specifications is essential. A slight change in their specified values may affect image quality, radiation dose and accuracy of measurements. Performance may also be influenced by the properties of the computer system and by major electromechanical components, that is, X-ray generator, X-Ray tube, detectors, collimators, alignment lights and patient's table. Operating conditions also influence the CT scanner's performance.

Material and Method

Development of an all purpose comprehensive CT phantom
CT phantoms are required for QA tests and performance evaluation of the CT scanners. An all purpose model has been designed considering the evolution of a comprehensive phantom which is very near to clinical situation. The various sections of the phantom has been prepared to correspond with the size and anatomical content of the human body. The insert materials used in the phantom have the same attenuation of the X-rays as in the human body. The equivalent materials used as inserts to represent various anatomical contents are in the density ranges of bone, fat, blood, soft tissue, body fluid, and body air cavities. The phantom encompasses all necessary features in most compact form for evaluating the performance standards, QA and dosimetry in CT scanner. It has been conceived for the first time and has been designed in complete conformity with the clinical needs. This phantom will be able to establish the traceability of performance and measurements of the CT scanner and also serve as a reference standard phantom which represents human body. The traceability of the results on similar make or different make CT scanner can be established using this phantom. This will also facilitate the calibration of QCT values of the different CT scanners against a standard phantom and the measurements will become traceable nationwide. It permits routine standardization of beam alignment, beam width, spatial uniformity, QCT, linearity, contrast, spatial resolution, line spread, noise, size independence tests and absorbed dose. All components of the phantom are housed in a compact transparent perspex tank which holds the system together in the desired orientation. A short list of QA parameters have been worked out for daily check-up of the CT scanner and patients clinical diagnostic evaluations have been performed only when the unit conformed with the standards set for quality image. This was done with a purpose to reduce the dosages.

Field uniformity and quantitative CT
It is most important to determine the spatial uniformity of CT number in a given media. This test has been conducted using water phantom for measuring the CT number of water, bone and air. In stimulated clinical conditions, calibration scan has been performed with narrowest slice width at all possible kVp stations. Average CT number in an region of interest (ROI) of approximately 1 cm^2 in the center of the phantom and at four locations at the phantom periphery are obtained (Table 1). The field view of the CT is completely uniform. Similarly using a

circular uniformity (water) phantom the aspect ratio (horizontal vs vertical diameter) has been measured and it has been found that this is in complete congruence of the actual value. This testified the caliper accuracy and reproducibility.

Table 1. Summary of Results of Quantitative CT Parameters

S. No.	Test	Measured	Remarks
1.	Water Phantom	CT number	± 3 CT no.
		Std. dev. 1	(tolerance)
2.	Body Phantom		
	(i) Bone	+ 995.5	± 3 CT no.
		Std. dev. 5.7	(tolerance)
	(ii) Air	–999	± 3 CT no.
		Std. dev. 3.5%	(tolerance)
3.	Horizontal vs vertical radii of a circle	Similar	Complete congruence
4.	Image documentation system	All measurement based on radiographic plan and with computer support were found in complete congruence	Complete reproducibility

Random Uncertainity in Pixel Value

A CT image of uniform object is both systematic and random variations in pixel numbers about some mean value. The random component of thin variation is pixel noise. Its effect on the image is to place a lower limit on the level of subject contrast which can be distinguished by the observer. Pixel noise is a critical limiting factor in CT since much soft tissue detail is low contrast in nature. The total pixel noise (N_p) is given by

$$N_p = \sqrt{N_e^2 + N_q^2}$$

Electronic noise (N_e) arises as random variation in detector signal prior to digitization and quantum noise (N_q) is due to random variation in the number of detected X-ray quanta. Electronic noise is thermal in origin and remains roughly constant. Quantum noise (N_q) arises from statistical uncertainty in the finite number of transmitted X – ray photons (n) collected in forming the image, i.e.

$$(N_q) \propto n^{-1/2}$$

and
$$N_q \propto 1/\sqrt{w^3 \cdot h \cdot Q}$$

w = spatial resolution element, h = image slice width, Q = tube mAs

Noise measurement is performed to assess the level of noise under stimulated clinical conditions and its variation with different scanning parameters. This is measured with the help of head and body water phantom and noise measurement

(N) is expressed as a per cent of the effective linear attenuation coefficient (μ_w) of water and corrected for the scanner contrast scale.

$$N = \sigma\, C_s \cdot 100/\mu_w$$

Where σ is the measured standard deviation of pixel values in the ROI and C_s is the contrast scale factor. C_s is defined as the change in effective linear attenuation coefficient for a given change in CT number for two known materials. Normally one material is water.

$$C_s = \mu_m(E) - \mu_w(E)/CT_m - CT_w$$

$\mu_m(E)$ and $\mu_w(E)$ are the linear attenuation coefficient of the reference material, and water, CT_m and CT_w is the CT number of material and water.

Measurement of Organ Dosages and Calibration of TLD

In this study entrance surface dose for the scan of skull, head and neck, thorax, abdomen, pelvis and spinal cord has been measured. TLD has been utilized for measuring primary X-ray contribution and scatter from different slices on adjoining areas by sticking multiple chips on the organs of interest. The TL chips were calibrated for various X-ray energies using Nucleonix TL reader. Phantom study for similar conditions for each site and X-ray parameters were carried out for standardization of the data. The TLD system maintained a SD of ± 5 percent. The TLD reader was checked daily for reproducibility by measuring a known dose and developing a correction factor for consistency and accuracy of the results. The depth dose data generated using TL chips in phantom at various depth was utilised for generating depth dose data which was utilized for organ dose estimation for various sites. The doses data thus generated has been projected in relation to site, organ, scatter dose, primary dose, kV, mAs, number of slices and field sizes etc. The organ doses data developed in this study is comparable to the EC countries.

TLD Dosimeter was utilized for measurement of organ dosages as it was found to be reproducible and will not perturb the clinical and anatomical evaluation. The standard deviation of the TL output was evaluated for the exposures on CT scanners to ascertain the reproducibility of results and finalise the acceptance/rejection criteria of output measurement.

The TL Dosimeter was calibrated for the known doses of Caesium-137 and Cobalt-60 to check the reproducibility. It was found that the TL dosimeter is able to produce the results within acceptable limits except few erratic results for gamma radiations.

The TL dosimeter was also calibrated for optimally utilized set of the following configurations (Table 2) of the X-ray parameters used in CT Scanners for scanning of the various organs.

Measurements for ESD were conducted for the desired anatomical sites/organs. The data has been collected in terms of exposure factors Vs. area per unit weight of the TL dosimeter output. The data has been converted in organ dose using the conversion factors developed based on our Phantom measurements.

Table 2. Calibration of TLD for Various CT Parameters

S. No.	Site Site	Kv	mAs (sec)	Time	Slice width (mm)	Wt (mg)	Glow Curve Area	Area/wt
1.	Skull	120	250	4.5	10.0	34.2	10769.90	314.91
2.	Maxilla	120	250	3.0	10.0	26.4	8484.82	321.39
3.	Thorax	120	300	3.0	10.0	16.2	4343.49	268.12
4.	Abdomen	120	300	3.0	10.0	15.6	4558.77	292.23
5.	Pelvis	120	300	3.0	10.0	16.2	4343.49	268.12
6.	Spine	120	300	3.0	05.0	15.6	5055.92	324.09

In this study organ dose measurement values has been utilized to optimize the image quality for reducing the patient dose. The data has been developed on Siemens, Hitachi and G.E. CT scanner to develop a National data bank. Some of the results obtained are listed in Table – 3

Table 3. Average Organ Doses in Common Examinations by Computed Tomography

Examination	Organ Receiving Dose	Dose from CT mGy
SKULL (top)	Brain	24.3–32.2
	Bone Marrow	3.1–4.1
	Thyroid	0.96–1.4
Thorax	Breast	23.5–27.1
	Lungs	20.0–25.5
	Bone Marrow	4.3–5.8
	Thyroid	2.3–3.1
Abdomen	Upper large intestine	20.0–26.7
	Lower large intestine	10.1–13.5
	Uterus	14.7–20.1
	Bone Marrow	6.9–9.7
Pelvis	Uterus	16.1–21.9
	Upper large intestine	7.5–10.3
	Lower large intestine	14.6–19.8
	Bone Marrow	5.6–7.9

Result and Discussion

The spatial resolution of CT scan is related to the entrance surface dose, therefore, for a quality image certain minimum X-ray dose could not be avoided. The repetition of CT scan due to malfunctioning and inaccuracy in the CT parameter has a very significant bearing on the dose reduction. The efficiency of X-ray detectors and image formation and its computation could alter the dosage required. Proper maintenance of QA parameters and daily QA check can reduce the dosage by choosing better scan parameters. The scan width and optimal number of CT cuts for required diagnosis may be the single largest factor effecting the dosage to the organ. The organ dosage data measured and reported in the study may serve as a guide line for choosing and planning the CT study for the organ of interest.

Acknowledgement

The authors express their profound gratitude to Sri Inder K. Khosla, Hony. Secretary and Treasurer, KNMH Society, New Delhi for his overwhelming encouragement for the study. The financial assistance provided by AERB for the evaluation is duly acknowledged.

References

1. Brooks, R.A. and G. Dichiro (1976) Statistical limitations in X-Ray reconstruction tomography. *Med. Phys.* 3: 237–240.
2. Gray, J.E. and N.T. Winkler, J.G. Srears, E.D. Frank (1992) *Quality Control in Diagnostic Imaging.* Rockville, MD: Aspen Systems, Inc.
3. Hanson, K.M. (1981) Noise and contrast discrimination in computed tomography. In *Radiology of the Skull and Brain: Technical Aspect of Computed Tomography.* T.H. Newton and D.G. Potts (Editors). The C.V. Mosby Co.
4. Judy, P.F., S. Batter, D.A. Bassano, E.C. Mc Cullough, J.T. Payne and Rothenberg (1977) Phantoms for Performance evaluations of CT Scanners AAPM report # 1: American Association of Physicists in Medicine.
5. Mishra, S.P. (1994) Quality assurance in C.T. scan-Our experience with Hitachi CTW 700. Invited talk delivered at international symposium. New Delhi: AIIMS.
6. Mishra, S.P., Vinita Gupta, S. Kumar, S. Khanduja and A.N. Vishnoi (1995) *Adequate Quality Assurance in Whole Body CT Scan—An Analysis in Indian Context.* Germany, Wurzberg: Roentgen Centenary Congress.
7. Specifications and acceptance testings of computed tomography scanners report of task group 2, Diagnostic X-Ray imaging committee, Published for the American Association of Physicists in Medicine AAPM Report No. 39.
8. Yester, M.V. and G.T. Barens (1977) Geometrical limitations of computed tomography (CT) scanner resolution. Proc. SPIE. Appl. Opt. Instr. in Medicine VI, 127: 296–303.

Radiobiology and Bio-Medical Research
Edited by K.P. Mishra
Copyright © 2004 Narosa Publishing House, New Delhi, India

14. Radiation, Glutathione and Cancer: Paradoxical Inter-relationship

V.N. Bhattathiri

Associate Professor of Radiotherapy and Head, Clinical Radiobiology Section
Regional Cancer Centre, Trivandrum

Abstract   Glutathione, a tripeptide with the multiple intracellular roles of aiding cell growth and protection against free radicals, can have paradoxical effect as both radioprotector and radiosensitizer as well as preventing and promoting cancer. If it is used without this knowledge in mind, one may end up with the effect opposite to that intended. The present article focuses on these paradoxical interrelationships.

Glutathione (GSH) is an ubiquitous tripeptide present in most cells, involved in diverse intracellular physiological and pathophysiological proceses, such as detoxification, defence against oxidative stress, protein and DNA synthesis and, along with other thiols, in protection of DNA from damage. But the multifaceted actions of GSH can result in paradoxical relationships in the context of radiation protection and sensitization, and cancer induction and progression. The present focuses on this paradoxicalness.

GSH as Radioprotector and Radiosensitizer

Glutathione affords radioprotection by radical scavenging, restoration of damaged molecules by hydrogen donation, reduction of peroxides and maintenance of protein thiols in the reduced state. Depletion of GSH, as well as protein thiols, by L-BSO, increases sensitivity of cells to radiation [11, 6]. Thiols other than GSH may be important in altering radiation response of cells. Cells with high levels of metallothioneins, rich in cysteine, have high levels of total sulphydryl groups and high radioresistance [3]. Glutathione and related enzymes form part of the system which protects cell membrane against radiation injury [28, 16, 7]. Increased levels of GSG and GST-pi are suggested as reasons for radioresistance cancers [25, 30]. Intravenous GSH helps ameliorate acute radiation effects [15]. Prior low dose radiation is known to induce radioresistance to subsequent higher doses. Experiments have shown that one reason for this is induction of denovo GSH synthesis or regeneration from GSS [18, 19, 20, 21]. Interestingly, this GSH induction affords protection not only against further radiation, but also various chemicals [19]. In the case of liver, elevation of GSH by previous low dose irradiation even affords protection against many chemicals which deplete GSH [23, 30]. It is interesting to consider the possibility that the reverse situation is also possible, that is, those who have higher GSH induced by exposure to other agents may have acquired resistance to radiation. There are many reports that chronic tobacco and alcohol habituates may have higher plasma and erythrocyte GSH content [32, 5]. It has been suggested [8] that acute tobacco smoke inhalation increases the oxidant burden on the lungs leading to transient depletion of GSH causing regulatory mechanism(s) in the lung to respond by increasing uptake of GSH precursors present in plasma disulfides. Evidently other organs rich in GSH, such as the liver, erythrocytes and kidney also respond to the oxidant insult by increased GSH synthesis and release into plasma. Elevated GST-pi has been reported in mucosa of chronic tobacco users and this suggests presence of an adaptive response [29]. Smokers thus may have either an elevated plasma GSH level or ability to respond efficiently and quickly to radiation induced GSH depletion which minimizes the risk of pneumonitis in them. In rats chronic ethanol consumption increases hepatic GSH, possibly due to stimulation following increased hepatic lipid peroxidation [2]. Johansson *et al.* [17] reported that in breast and oesophageal cancer patients treated by radiotherapy, the occurrence of radiation pneumonitis was almost exclusively limited to non-smokers. One interesting possibility here is that individuals with such chronic exposure to chemicals or radiation and with high GSH, may be better suited for work related to nuclear accidents; a 'volunteer' army may be formed from among them to cope with nuclear emergencies.

In a study by the author it was observed that even though those with higher tumor GSH in relation to plasma GSH had aggressive and more rapidly proliferating tumors [5]. Yet, such tumors had better control with radiotherapy, provided the treatment was sufficiently intense. The reason is that radiation induces mitotic cell death, and the greater the proliferation rate, greater the death, provided the radiation schedule is adequate. This also is the reason for the greater sensitivity of rapidly proliferating tissues such as the germ cells, intestinal mucosa and bone marrow. It is possible that once radiation exposure has already occurred, decreased availability of GSH may decrease the growth rate of various cell systems and thus decrease the normal tissue damage. This is an avenue of research for potential application for nuclear casualty management.

Thus, it is seen that GSH acts as a radiation protector, due to its free radical scavenging activity, when it is present in sufficient quantities at the time of radiation. But, once the radiation exposure is over and damage has already occurred, GSH increase may accelerate the growth rate and precipitate mitosis induced cell death. It is not known whether this concept can be put to any use for the purpose of protection in times of whole body exposure. The world is on the look out for a substance or group of substances that can be used at the time of nuclear emergencies, and GSH fits the bill. But inappropriate use may lead to deleterious effect. This also applies to putative radioprotectors that elevate GSH levels.

GSH in Cancer Prevention and Promotion

Ever since the identification of free radical mediated oxidative injury as the root cause of cancer induction, the world has been on the look out for a substance or group of substances that can protect cells from this and prevent cancer. Many substances such as retinoids, vitamin E and those selenium, have been propounded as having this property. Considering its intracellular role, GSH and its inducers are also prime candidates for this. GSH- and GSH-dependent enzymes play a crucial role in tobacco-related tumourigenesis, and has potential to be used as markers of carcinogen exposure [31]. Interindividual variation in risk for bronchogenic carcinoma results in part from variation in expression of GSH related antioxidant genes in bronchial mucosal cells [9]. It is also known that SH groups other than that of GSH may be important in preventing carcinogenesis. N-acetyl cysteine, a precursor of intracellular glutathione, can prevent chemically induced cancers in animals [10]. It has been reported that erythrocyte glutathione peroxidase was depressed in head and neck of cancer patients [14]. It has been suggested that the chemopreventive effect of S-allylcysteine is through enhanced production of GSH and related enzymes [4]. Topical application of GSH on cutaneous papillomas inhibited development of malignancy in them in a dose dependent manner and application of buthionine sulfoximine (BSO), an inhibitor of GSH synthesis, enhanced tumor progression [27].

GSH is intimately involved in protein and DNA synthesis, substances which are necessary for division and growth of all cells, including cancer cells, suggesting it may promote tumor growth. GSH administration to rats given carcinogens resulted in an increase in hepatoma, implying that GSH may stimulate progression of cancer [1]. Dramatic increases in activity of gamma-glutamyl transpeptidase (GGT), an enzyme that is intimately concerned in the synthesis and metabolism of glutathione, are found in many chemically induced animal tumours, and can be recognized in pre-malignant cells long before any morphological changes become evident implying that GSH supply to persons who already have tumors at this stage of development may have enhanced tumor growth [13]. Glutamine, one of the raw materials for GSH synthesis often administered to cancer patients to help ameliorate cancer cachexia and prevent gut and oral toxic side-effects of radiotherapy, may also stimulate tumor growth [24]. Cell-culture studies have shown that depletion of intracellular GSH by BSO can have cytotoxic effect against tumor cells [12]. It is reported that ethanol fed hamsters develop DMBA induced buccal pouch cancers at greater frequency and this is preceeded by a fall in GSH content, and later a rise at the time of occurrence of the tumors (Nachiappan *et al* 1993). It is logical to assume that the tumors were induced at the time of fall in GSH level, and the later rebound rise in GSH level could have contributed to their growth. In a large study it was observed that patients given N-acetyl cysteine had higher incidence of second primary tumors [33].

Thus it is seen that GSH can either prevent or promote cancer depending on whether there are transformed cells present in the body. Inappropriate GSH administration can lead to effect opposite to that desired.

References

1. Ahluwalia, M.B., J. Rotstein, M. Tatematsu, M.W. Roomi and E. Farber (1983)

Failure of glutathione to prevent liver cancer development in rats initiated with diethylnitrosamine in the resistant hepatocyte model. *Carcinogenesis.* 4: 119–21.

2. Aykac, G., M. Uysal, A.S. Yalcin, N. Kocak-Toker, A. Sivas, H. Oz (1985) The effect of chronic ethanol ingestion on hepatic lipid peroxide, glutathione, glutathione peroxidase and glutathione transferase in rats. *Toxicology* 36: 71–6.

3. Bakka, A., A.S. Johnsen, L. Endresen and H.E. Rugstad (1982) Radioresistance in cells with high content of metallothionein. *Experientia* 38: 381–3.

4. Balasenthil, S. and S. Nagini (2000) Inhibition of 7, 12-dimethylbenz [a] anthracene-induced hamster buccal pouch carcinogenesis by S-allylcysteine. *Oral. Oncol* 36: 382–386.

5. Bhattathiri, V.N. (1997) Glutathione and Cancer: Present status and future prospects. In: *Radiation, Radiomodifiers and Human Health.* Devi, P.U., Bisht, K.S. and Rao, B.S.S. Editors National Institute of Science Communication. New Delhi: CSIR.

6. Biaglow, J.E., M.E. Varnes, S.W. Tuttle, N.L. Oleinick, K. Glazier, E.P. Clark, E.R. Epp. and L.A. Dethlefsen (1986) The effect of L-buthionine sulfoximine on the aerobic radiation response of A549 human lung carcinoma cells. *Int. J. Radiat. Oncol. Biol. Phys*: 12: 1139–42.

7. Clark, E.P., E.R. Epp, M. MorseGaudio and J.E. Biaglow (1986) The role of glutathione in the aerobic radioresponse. I Sensitization and recovery in the absence of intracellular glutathione. *Radiat. Res* 108: 238–50.

8. Cotgreave, I.A., U. Johansson, P. Moldeus and R. Brattsand (1987) The effect of acute cigarette smoke inhalation on pulmonary and systemic cysteine and glutathione redox states in the rat. *Toxicology* 45: 203–212.

9. Crawford, E.L., S.A. Khuder, S.J. Durham, M. Frampton, M. Utell, W.G. Thilly, D.A. Weaver, W.J. Ferencak, C.A. Jennings, J.R. Hammersley, D.A. Olson, J.C. Willey (2000) Normal bronchial epithelial cell expression of glutathione transferase Pl, glutathione transferase M3, and glutathione peroxidase is low in subjects with bronchogenic carcinoma. *Cancer Res* 60: 1609–18.

10. De Flora, S., M. Easting, D. Serra and C. Bennicelli (1986) Inhibition of urethane-induced lung tumours in mice by dietary N-acetylcysteine. *Cancer Lett.* 32: 235–41.

11. Dethmers, J.K. and A. Meister (1981) Glutathione export by human lymphoid cells: depletion of glutathione by inhibition of its synthesis decreases export and increases sensitivity to irradiation. *Proc. Natl. Acad. Sci. USA* 78: 7492–6.

12. Dorr, R.T., J.D. Liddil and M.J. Soble (1986) Cytotoxic effects of glutathione synthesis inhibition by L-buthionine-(SR)-sulfoximine on human and murine tumor cells. Invest. *New Drugs* 4: 305–13.

13. Goldberg, D.M., Structural (1980) Functional, and clinical spects of gamma-glutamyl transpeptidase. *CRC Crit. Rev. Clin. Lab. Sci* 12: 1–58.

14. Goodwin, W.J. Jr., H.W. Lane, K. Bradford, M.V. Marshall, A.C. Griffin, H. Geopfert and R.H. Jesse (1983) Selenium and glutathione peroxidase levels in patients with epidermoid carcinoma of the oral cavity and oropharynx. *Cancer* 51: 110–5.

15. Grozdov, S.P. (1987) Effect of glutathione on the course of radiation sickness with a marked gastrointestinal syndrome and the effectiveness of parenteral feeding in an experiment. *Radiobiologiia* 27: 657–662.

16. Helszer, Z., Z. Jozwiak and W. Leyko (1980) Osmotic fragility and lipid peroxidation of irradiated erythrocytes in the presence of radioprotectors. *Experientia* 36: 521–4.

17. Johansson, S., L. Bjermer, L. Franzen and R. Henriksson (1998) Effects of ongoing

smoking on the development of radiation induced pneumonitis in breast cancer and oesophageal cancer patients. *Radiother. Oncol* 49: 41–47.

18. Kojima, S., O. Matsuki, I. Kinoshita, T.V. Gonzalez, N. Shimura and A. Kubodera (1997) Does small-dose gamma-ray radiation induce endogenous antioxidant potential in vivo? *Biol. Pharm. Bull.* 20: 601–4.

19. Kojima, S., O. Matsuki, T. Nomura, A. Kubodera and K. Yamaoka (1998) Elevation of mouse liver glutathione level by low-dose gamma-ray irradiation and its effect on CC14-induced liver damage. *Anticancer Res.* 18: 2471–6.

20. Kojima, S., O. Matsuki, T. Nomura, A. Kubodera, Y. Honda, S. Honda, H. Tanooka, H. Wakasugi and K. Yamaoka (1998) Induction of mRNAs for glutathione synthesis-related proteins in mouse liver by low doses of gamma-rays. *Biochim. Biophys. Acta.* 1381: 312–8.

21. Kojima, S., O. Matsuki, T. Nomura, N. Shimura, A. Kubodera, K. Yamaoka, H. Tanooka, H. Wakasugi, Y. Honda, S. Honda and T. Sasaki (1998) Localization of glutathione and induction of glutathione synthesis-related proteins in mouse brain by low doses of gamma-rays. *Brain Res.* 808: 262–9.

22. Kojima, S., S. Matsumori, H. Ono and K. Yamaoka (1999) Elevation of glutathione in RAW 264.7 cells by low-dose gamma-ray irradiation and its responsibility for the appearance of radioresistance. *Anticancer Res.* 19: 5271–5.

23. Kojima, S., H. Shimomura, S. Matsumori (2000) Effect of pre-irradiation with low-dose gamma-rays on chemically induced hepatotoxicity and glutathione depletion. Anticancer Res. 20: 1583–8.

24. Miller, A.L. (1999) Therapeutic considerations of L-glutamine: a review of the literature. *Altern Med. Rev.* 4: 239–48.

25. Mulder, T.P., J.J. Manni, H.M. Roelofs, W.H. Peters, A. Wiersma (1995) Glutathione S-transferases and glutathione in human head and neck cancer. *Carcinogenesis* 16: 619–24.

26. Nachiappan, V., S.I. Mufti, C.D. Eskelson (1993) Ethanol-mediated promotion of oral carcinogenesis in hamsters: association with lipid peroxidation. *Nutr. Cancer* 20: 293–302.

27. Rotstein, J.B. and T.J. Slaga (1988) Effect of exogenous glutathione on tumor progression in the murine skin multistage carcinogenesis model. *Carcinogenesis* 9: 1547–51.

28. Ryskulova, S.T., V.P. Verbolovich, E.P. Petrenko, T.V. Tsvetkova and T.A. Balakhchi (1983) Antiradical protective systems in the plasma membranes of an irradiated animal. Radiobiologiia. 23: 648–50.

29. Sarkar, G., N. Nath, N.K. Shukla, R. Ralhan (1997) Glutathione S-transferase pi expression in matched human normal and malignant oral mucosa. *Oral Oncol.* 33: 74–81.

30. Sarkar, S.R., L.R. Singh, B.P. Uniyal and B.N. Chaudhuri (1983) Effect of whole body gamma radiation on reduced glutathione contents of rat tissues. *Strahlentherapie* 159: 32–3.

31. Saroja, M., S. Balasenthil and S. Nagini (1999) Tissue lipid peroxidation and glutathione-dependent enzyme status in patients with oral squamous cell carcinoma. *Cell Biochem. Funct.* 17: 213–6.

32. Sinues, B., P. Rueda, M.A. Saenz., M.L. Bernal and A. Alcala (1990) Erythrocyte glutathione and urinary thioethers in smokers. *Med. Clin. (Barc.)* 95: 725–727.

33. Van Zandwijk, N., O.Dalesio, U. Pastorino, N. de Vries, H. van Tinteren (2000) EUROSCAN, a randomized trial of vitamin A and N-acetylcysteine in patients with head and neck cancer or lung cancer. For the European Organization for Research and Treatment of Cancer Head and Neck and Lung Cancer Cooperative Groups. *J. Natl. Cancer Inst.* 21: 977–86.

Radiobiology and Bio-Medical Research
Edited by K.P. Mishra
Copyright © 2004 Narosa Publishing House, New Delhi, India

15. Involvement of Signal Transduction Pathways in Mustard Gas Induced Lung Injury

Diptendu Chatterjee, Shyamali Mukherjee and Salil K. Das

Department of Biochemistry, Meharry Medical College
1005 D.B. Todd Blvd, Nashville, TN 37210

Abstract: The exact mechanism by which mustard gas exposure causes lung injury including ARDS is not well known. Since human volunteers can not be used for this type of study, we have developed a guinea pig model. Our initial studies have shown that guinea pigs exposed to half mustard gas intratracheally, accumulate high levels of TNF-alpha, followed by activation of both acidic and neutral sphingomyelinases resulting in high accumulation of ceramides, a second messenger involved in cell apoptosis. This signal transduction event was associated with alteration in oxygen defense system and accumulation of free radicals. These biochemical changes were associated with lung damage, including edema, congestion, hemorrhage and inflamation and exclusion of type II cells into alveolar space. It is important to note that there was an activation of NFkB for a short period (one to two hours after exposure); however, NFkB rapidly disappeared after two hours. It is possible that the initial activation was due to an adaptive response to protect the cells from damage since it is known that NFkB is an inhibitor of TNF-alpha/ceramide induced cell apoptosis. Since NFkB disappeared after two hours, the cells continued being damaged due to accumulation of ceramides.

Introduction

Sulfur mustard gas (SMG) has been used as a vesicant chemical warfare agent. Inhalation of SMG causes hemorrhagic inflammation to the tracheobronchial tree with severe pulmonary complications including Adult Respiratory Distress Syndrome (ARDS) [11]. Most deaths are due to secondary respiratory infections. Several studies in rats and mice have shown that the mechanism of SMG action on lung, skin or other organs includes DNA alkylation, cross linking of DNA [6], activation of proteases [4], free radicals production [6], inflammations [9] and activation of tumor necrosis factor (TNF α), a part of the inflammatory cytokine cascade [1]. Among the numerous mechanism proposed for the SMG induced ARDS, it appears that the initiation of TNFα cascade is the major pathway. Our ultimate goal is to develop drugs to counteract the SMG induced lung injury.

Since human volunteers cannot be used for such studies, we need to develop a good animal model for the study of SMG toxicity. We have established earlier that structurally and functionally, guinea pig lungs are more alike to human lungs in comparison to other animal species. Therefore, we have studied here mechanisms by which a version of SMG, 2-chloroethyl ethyl sulfide [Cl–CH$_2$CH$_2$–S–CH$_2$CH$_3$, CEES], referred as 'Half Mustard Gas', HMG causes lung injury. It will be helpful for us to design drugs to prevent SMG induced lung injury after getting the clear picture of the SMG action on guinea pig lung.

Materials and Methods

Chemicals
Half mustard gas, 2-chloroethyl ethyl sulfide [Cl–CH$_2$CH$_2$–S–CH$_2$CH$_3$, CEES], was obtained from Sigma Chemicals (St. Louis, MO). TNF-αELISA kit was obtained from BioSource International (Camarillo, CA).[14] C-Sphingomyelin (50 mCi per mmol) was obtained from American Radiolabeled Chemicals, Inc. (St. Louis, MO), Diacyglycerol kinase kit and ^{32}P-γ-ATP was obtained from Amersham Pharmacia Biotech, Piscataway, NJ.

Injection of mustard gas in guinea pigs
Male guinea pigs (Hartley strain, five to six weeks old, 400 gm body weight) were obtained from Harlan Sprague Dawley Inc (Indianapolis, Indiana). Animals were injected intratracheally single doses of HMG (0.5 mg/kg body weight) in ethanol (total injection volume was 100 μl per animal) Control animals were injected with 100 μl of ethanol in the same way intratracheally. The animals were sacrificed at different time intervals after HMG injection and lung was lavaged as described before [8]. Lavage macrophages were obtained from lavaged fluid by centrifugation at 600 xg for 10 minutes. The cells, tissue and lavage fluid were kept at –70°C for future use.

Assay of TNF-α
Cells or tissue were sonicated in Tris buffered saline, pH 7.4, centrifuged at 10,000 rpm for 10 minutes. TNF-α was measured in the supernatant using TNF-α ELISA kit. Supernatants were added to TNF-α antibody coated micro wells along with biotin conjugated second antibody and incubated for an hour at room temperature. After washing, the wells were incubated with streptovadin-horse radish peroxidase complex. The wells were washed and color reactions were carried out using horse radish peroxidase substrate. The color developments were measured using ELISA reader after addition of stop solution.

Assay of sphingomyelinase
Both acid and neutral sphingomyelinases were assayed using ^{14}C-sphingomyelin as described by Gulbins and Kilesnick [7]. Briefly, tissue homogenates in neutral (pH7.4) or acidic (pH 5.0) buffer were centrifuged at 1,000 rpm for 5 minutes to remove nuclear fraction. The supernatants were incubated with ^{14}C-sphingomyelin in acidic (for acid sphingomyelinase assay) or neutral buffer (for neutral sphingomyelinase assay) for 30 minutes. Following enzyme assay the products were separated by chloroform: methanol: water (2:1:1) extraction. Radioactivity of the upper aqueous layer gave a measure of the sphingomyelinase activity.

Assay of ceramide
Ceramides were assayed in total lipid from lung tissue according to Cao *et al* [3] using diacylglycerol kinase (DAG kinase) kit (Amersham Pharmacia Biotech, Piscataway, New Jersey). Total lipids were extracted from lung tissue by chloroform: methanol (2:1) extraction and dried under nitrogen. Dried lipids were dissolved in imidazole buffer containing n-octyl β-glucopyranoside and DAG kinase. The reaction was initiated by addition of ^{32}P-γ-ATP. Following reaction, the products were separated by TLC and autoradiographed to visualize ceramide level. Ceramide standard was run side by side to identify ceramides on TLC. The ceramide containing portions of the TLC were scraped out and counted in scintillation counter after addition of scintillation fluid to estimate ceramide content quantitatively.

Assay of NFkB
Activation of NFkB was measured in the nuclear extract from control and HMG exposed lung tissues according to Mackay *et al* [10] by gel shift assay. Nuclear fractions were isolated by low speed centrifugation and extracted with high salt buffer. The nuclear extracts were incubated with ^{32}P-γ-ATP labeled oligo nucleotide probe for NFkB (NFkB in its activated form binds specifically to this oligo nucleotide probe). The reaction mixtures were separated by native polyacrylamide gel electrophoresis and autoradiographed to visualize the activation of NFkB.

Light and electron microscopy study
Morphological examinations were carried out as described before [5]. Briefly, a portion of lung tissue from control and HMG exposed animals were fixed over night in neutral buffered formalin, pH 7.5, and embeded in paraffin for light microscopy. Five microns sections were stained with hematoxylin and eosin stain. For electron microscopy studies the tissues were fixed over night in 4 percent glutaraldehyde-paraformaldehyde and embed in epon. One micron sections were examined under an electron microscope after staining with uranyl acetate and lead citrate.

Results

Effects of mustard gas exposure on TNF-α level in lungs
We have given single intratracheal injection (0.5 mg/kg body weight) of HMG to guinea pigs. At different time points, guinea pigs were sacrificed and lung was removed after perfusion. The lung was lavaged and TNF α concentrations were measured in lung lavage fluid, lung lavage macrophages and in lung tissue. The results indicated that the level of TNF α in lavage fluid was very low where as a high level of TNF α accumulated in lung as well as in lavage macrophage within an hour of HMG exposure. The level of TNF α decreased rapidly after an hour and came to the normal level within 24 hours of HMG exposure (Fig. 1).

Effects of mustard gas exposure on sphingomyelinase levels in lungs
The results of this study clearly demonstrated that both the acidic and neutral sphingomylinase activities were increased four to five fold after HMG treatment.

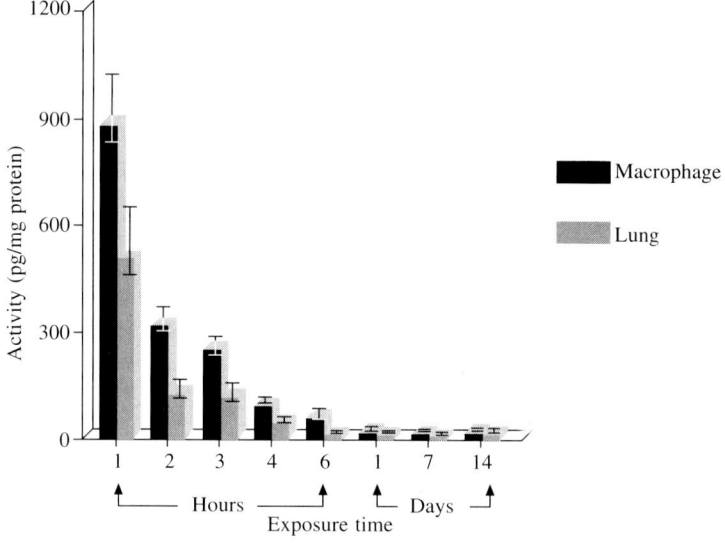

Figure 1. Induction of TNF-α in guinea pig lung after exposure to half mustard gas.

As the control level of acidic sphingomylinase activity was much higher than the control level of neutral sphingomylinase, we could see much higher activity of acid sphingomylinase compared to neutral sphingomylinase after HMG treatment (Figs. 2-3). Both acid and neutral sphingomyelinase activities started to increase along with the increase of TNF α but gave a maximum peak within four to six

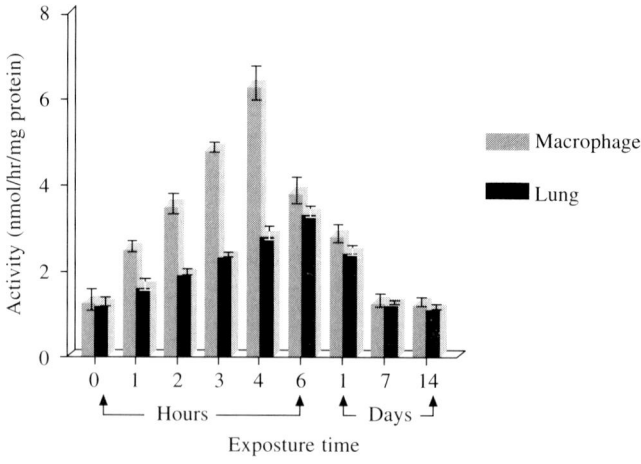

Figure 2. Activation of neutral sphingomyelinase after exposure of HMG in guinea pig lung macrophage and lung tissue.

hours after HMG exposure. The level of sphingomyelinase decreased rapidly to come back close to normal level within 24 hours.

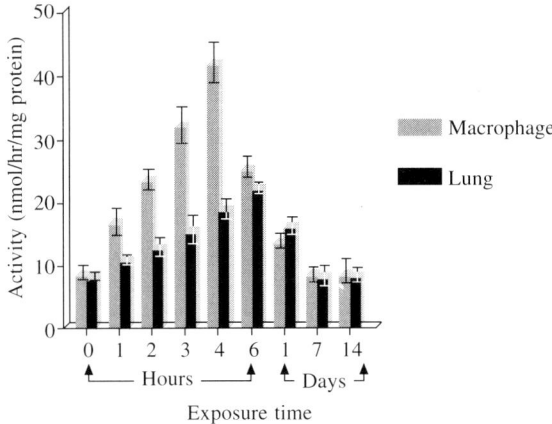

Figure 3. **Activation of acid sphingomyelinase after exposure of HMG in guinea pig lung macrophage and lung tissue.**

Effects of mustard gas exposure on ceramide levels in lungs
The ceramides accumulation followed by HMG exposure demonstrated a biphasic pattern. Within an hour of HMG exposure, ceremides level became very high and gave a peak accumulation within two hours (Figs. 4-5). After two hours, there was some decrease in the ceramides level but again the level increased to a very high level and remained almost steady even upto 14 days (slight decrease in ceramides level at day 14).

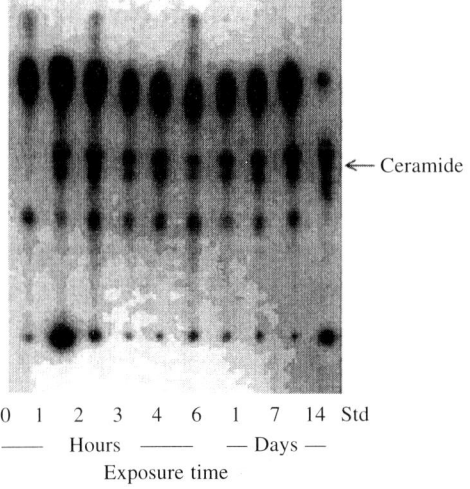

Figure 4. **Effects of HMG exposure on the accumulation of ceramides in lung tissue as visualized by autoradiography after separation of the reaction mixture by TLC.**

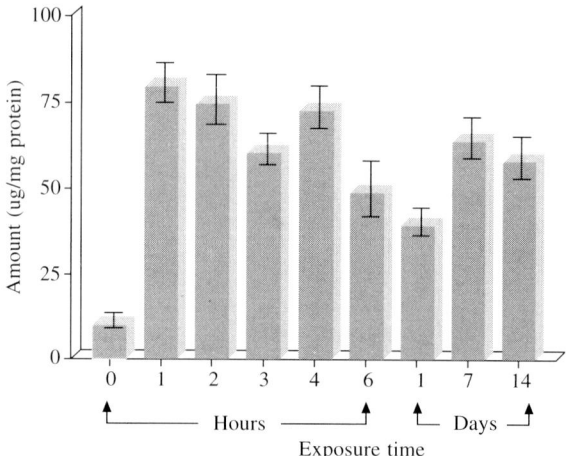

Figure 5. Quantitative estimation of ceramide accumulation after exposure of HMG in lung tissue.

Effects of mustard gas exposure on NFkB levels in lungs

The NFkB activation was measured in the nuclear extracts of lungs after exposure to HMG. NFkB showed activation only upto two hours after HMG exposure (Fig. 6). After two hours of HMG exposure little or no activated NFkB could be observed.

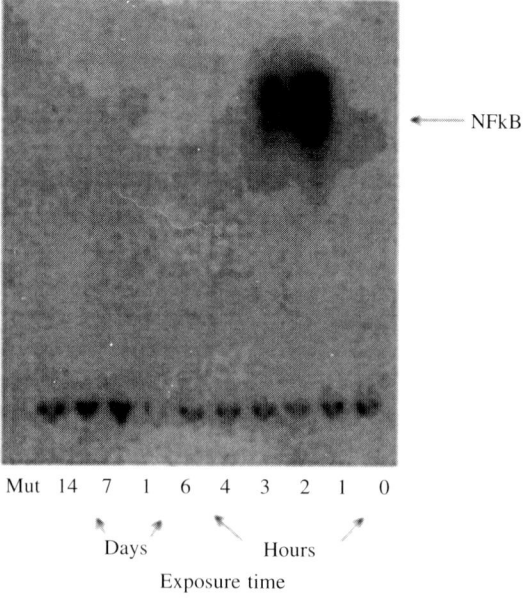

Figure 6. Activation of NFkB as observed by gel shift assay of the nuclear extracts of lung tissues after exposure to the HMG.

Histological and ultrastructural studies of lung after HMG exposure

Light microscopy studies showed evidence of severe lung damage, including edema, congestion, hemorrhage and inflammation after exposure to HMG. (Fig. 7 A and 7B). The electron microscopic studies demonstrated damage of Type II cells and spillage of alveolar Type II cells in alveolar space (Fig. 8A and 8B).

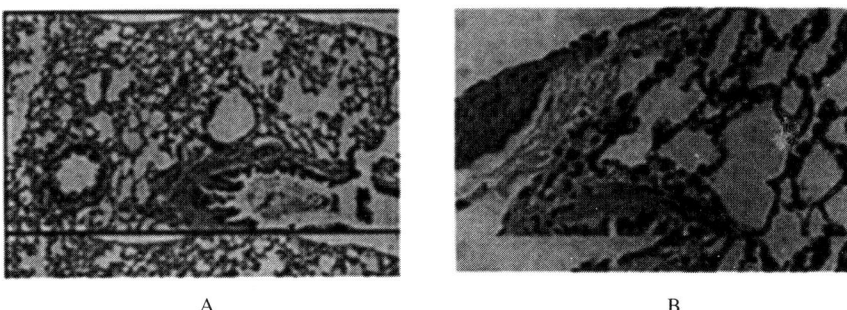

A   B

Figure 7.   Light microscopy study of the lungs after two hours exposure to the HMG. (A) X 100 and (B) X 200

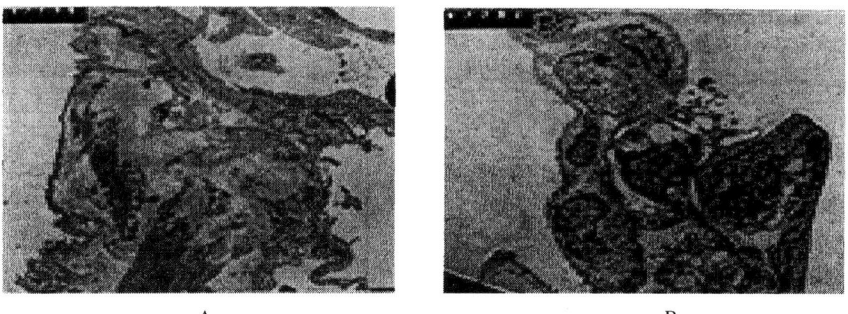

A   B

Figure 8.   Electron microscopy study of the lungs after exposure to the HMG. (A) one hour exposure; and (B) two hours exposure

Discussion

We have developed the guinea pig model to study mustard gas induced lung injury. Guinea pigs are closer to humans than other primates so far as structure and function of lungs are concerned, and it is difficult to get human volunteers a guinea pig model will be ideal for studying the mechanism of lung injury by mustard gas.

After intratracheal injection of HMG to guinea pigs, the TNF-α level increased sharply within an hour of exposure. TNF-α level started declining after an hour and came back to basal level within 24 hours. The stimulation of sphingomyelinase activity followed the accumulation of TNF-α giving a peak within four to six hours after HMG exposure. Both the acid and neutral sphingomyelinase activities were stimulated but the level of acid sphingomyelinase was found to be much higher after HMG exposure as the control level of acid sphingomyelinase activity was much higher than the control level of neutral sphingomyelinase. As the sphingomyelinase activity increased, the accumulation of ceramides started.

Ceramide level increased within one hour of HMG exposure and after a slight fall at three to six hours, it increased again and remained at high level even upto 14 days after HMG exposure. The activation of NFkB coincided with the increase of TNF-α in lung tissue. NFkB is well known to oppose the TNF-α mediated apoptosis [11]. Hence, we could see a biphasic effect of HMG on lung. After initial damge by TNF-α within two hours there was some recovery due to activation of NFkB. Slight drop in the ceramide level between three to six hours might be due to the sharp transitional activation of NFkB at one to two hour after HMG exposure.

Acknowledgement

This work was supported by grants from the US Department of Army (DAMD-17-99-1-9550) and the National Institutes of Health (2S06-GM-08037).

References

1. Arroyo, C.M., R.L. Von Tersch and C.A. Broomfield (1995) Activation of alpha=human tumor necrosis factor (TNF-α) by human monocytes (THP-1) exposed to 2-chloroethyl ethyl sulfide (H-MG). Human and Experimental. *Toxicology* 14: 547–553.
2. Beg, A.A. and D. Baltimore (1996) An essential role for NFkB in preventing TNF α induced cell death. *Science* 274: 782–787.
3. Cao, L.C., T. Honeyman, J. Jonassen and C. Scheid (2000) Oxalate-induced ceramide accumulation in Madin-Darby canine kidney and LLC-PK1 cells. *Kidney International* 57: 2403–2411.
4. Cowan, F.M. and C.A. Broomfield (1993) Putative roles of inflamation in the dermatopathology of sulfur mustard. *Cell Biol. and Toxicol.* 9: 201–213.
5. Das, S.K., S. Mukherjee and U. Desai (1994) Development of pancellular toxicity in guinea pig lung by ingestion of oleylanilide. *J. Biochem. Toxicology* 9: 41–49.
6. Elsayed, N.M., S.T. Omaye, G.J. Klain and D.W. Korte (1992) Free radical-mediated lung response tp the monofunctional sulfur mustard butyl 2-chloroethyl sulfide after subcutaneous injection. *Toxicology*, 72: 153–165.
7. Gulbins, E. and R. Kolesnick (2000) Measurement of sphingomyelinase activity. *Methods in Enzymology* 322: 382–388.
8. Khan, A.Q., O.S. Mathew and S.K. Das (1985) Phospholipid composition of guinea pig lung lavage. *Lipids* 20: 7–10.
9. Kuhns, D.B., E. Decarlo, D.M. Hawk and J.I. Gallin (1992) Dynamics of the cellular and humoral components of the inflamatory response elicited in skin blisters in humans. *J. Clin. Invest.* 89: 1734–1740.
10. Mackay, F., G.R. Majeau, P.S. Hochman and J.L. Browning (1996) Lymphotoxin β receptor triggering induces activation of the nuclear factor kB transcription factor in some cell type. *J. Biol. Chem.* 271: 24934–24938.
11. Momeni, A.Z., S. Enshaelh, M. Meghdadi and M. Amindjavaheri (1992) Skin manifestation of mustard gas. Clinical study of 535 patients exposed to mustard gas. *Archs. Dermatol.* 128: 775–780.

New Advances in Bio-Medical Research

Radiobiology and Bio-Medical Research
Edited by K.P. Mishra
Copyright © 2004 Narosa Publishing House, New Delhi, India

16. Biosafety: Problems and Prospects

D. Subrahmanyam

Ace Diagnostics and Biotechnology Limited, Plot. No. 66, Sector 18, HUDA,
Gurgaon, Haryana 121001, India

Abstract: Advances in modern biotechnology are increasingly applied to product development for improved health-care, agricultural outputs and environment. The biotechnology industry has become a multi-billion dollar industry. The extensive use of rDNA technology in industry often entails release of genetically modified organisms (GMOs) into the environment. There is a global concern that uncontrolled release of GMOs may bring about adverse effects on environment and health of humans and animals. There is a strong support among the nations that internationally binding biosafety regulations are needed for orderly application of biotechnology that ensures safety to health and environment. The national and international initiatives towards this goal have been outlined and the need for capacity strengthening of developing countries in this area are emphasized.

Introduction

The advent of modern biotechnology, particularly involving application of genetic engineering over the last two decades, has brought an industrial revolution and resulted in numerous high value-added products that had great impact on health, agriculture and environment. Thus biotechnology industry has become a multi-billion dollar venture and a driving economic force. The enormous thrust made by the industry using this technology brought in its wake concerns regarding safety to health of humans and animals and to the environment. This presentation deals with a brief review of the advances occurring in this area; their potential impact on health, agriculture and environment; and the global approaches being made to tackle the biosafety issues under the United Nations System.

Applications of Biotechnology

In the area of health-care application, modern biotechnology lead to sequencing of human genome and a better knowledge of disease processes. The sequence databases held in gene banks together with advances in bioinformatics are likely to change the face of biomedicine. Application of recombinant gene technology facilitated rapid production of vital pharmaceuticals, diagnostics, vaccines and opened new vistas in effectively tackling genetic disorders. Products such as humulin, erythropoietin, tissue plasminogen activator (TPA), human growth hormone, several mediators and clotting factors have already been in the market.

A new field of pharmacogenomics has arisen which promises to yield medicines tailored to specific altered genes. Recombinant antigens, polymerase chain reaction (PCR) applications, DNA probes and monoclonal antibodies have revolutionized diagnostic technology. Advances in genetic engineering resulted in improvement of the existing vaccines, elimination of several vaccine-related untoward reactions, development of single dose delivery of multiple protective epitopes, slow release formulations, DNA vaccines, and efficient vectors for vaccine delivery. More recently plant-based edible vaccines are being developed which offer great promise in combating infections ranging from bacterial to viral infections [3].

Application of modern biotechnology to agriculture has added a new dimension in our ability to meet the ever-increasing needs for food consequent to population explosion [6]. Cash crops that are resistant to pests such as viruses, insects and fungi have been developed and field-tested. Plants, which are herbicide resistant and tolerant to abiotic stress, are being cultivated. Genetically modified crop plants, which yield products with increased shelf life and improved quality and nutritional value, have been developed and marketed.

Modern biotechnology is also being used for cleaner environmental applications. Biopesticides, biofertilizers, biodegradable plastics and bioleaching are replacing the conventional polluting processes. The technology is increasingly applied for waste disposal and management; waste water purification and bioremediation.

Concerns

The advent of rDNA technology provided unpreceedented opportunities for horizontal and vertical transfer of genetic elements which brought about serious concerns of safety to human and animal health and to the environment in lay public, environmentalists, decision makers, and in even among some leading scientists. Application of genetically modified organisms (GMOs) for industrial development often entails release of the GMOs into the environment. Several releases have already taken place in not only developed countries but also in developing countries of Asia and the Far East and in Latin America. Considerable global concerns have been expressed that such a release may result in deleterious effects such as colonization of non-indigenous organisms, disruption of ecosystems [5], and inadvertent evolution of pathogenic traits. It is pertinent to record here few examples of such actual and perceived risks.

1. Potential exists for out-crossing to a wild relative species that might coexist with the transgenic crop creating super weeds [2]. Monsanto decided not to pursue the development of herbicide-resistant sorghum that out-crosses with a weed (Johnson grass). In a recent study Wilkinson from the UK. observed that herbicide resistance could spread from genetically modified rapeseed oil to its wild relatives like 'wild turnip'.
2. Vector-mediated horizontal gene transfer or vector recombination may generate new pathogenic bacteria or new virulent strains of virus [7]. A gene to resist a herbicide introduced into rapeseed was found in bacteria isolated from the gut of bees that ate pollen from rapeseed.
3. Modified plants may contain novel proteins that may provoke allergic

reactions. Recently Aventis, USA has withdrawn Taco shells from the market as it was contaminated with a Bt-protein Cry9C, a potential allergen, which was not approved for human consumption.

4. Extensive use of Bt-toxin containing crops can have potentially negative impact on ecological processes and non-target organisms. Pollen from maize engineered to make Bt-insecticide could kill monarch butterflies thus causing ecological imbalance. Heavy herbicide applications in the field result in severe reduction in weeds that eventually affect the bird population that live on them.

5. Insect pests develop resistance to crops with Bt-toxins.

6. Widespread use of a single transgenic cultivar may lead to a loss of genetic diversity.

7. Rats fed on genetically modified potatoes developed defective immune responses.

8. Developing countries have also expressed socio-economic and ethical concerns such as, biotechnology products substituting their agricultural and industrial exports; inadequate expertise in adoption and practice of new technologies and increazed marginalization of small farmers and fall in employment in their countries.

While the government and industry in the USA feel products made through genetic engineering are practically no different from similar ones made by conventional methods, those in Europe, Japan, Africa and some countries of Asia and the Far East are apprehensive in their commercialization.

Risk-Benefit Analysis

There seems to be a need for a careful analysis of risks, both hypothetical and real in application of biotechnology for industrial and economic development [4] However, the key contributions of biotechnology and risks of not availing its benefits such as loss of rain forests, low food output and consequent hunger and malnutrition in expanding population, inadequate new tools to fight diseases and environmental degradation must be addressed before taking any steps to limit its use. It should be realized that there is nothing like zero risk in any new technology. One should consider what is acceptable risk in relation to the benefits seen in the application of the technology. Risk analysis should be done step-by-step and case-by-case and should be based on sound scientific principles and on the quality of the product rather than the process adopted.

International Initiatives on Biosafety

United Nations Industrial Development Organization (UNIDO) at Vienna is one of the earliest international agencies, which realized the potential of the modern biotechnology, and the benefits that could be derived on its application for industrial development. Having been actively involved since the early 1970s in fostering biotechnology in developing countries, UNIDO realized the need for orderly application of biotechnology that ensures safety to health and environment. To this end UNIDO set itself certain objectives to promote internationally acceptable biosafety policies and guidelines, training and providing technical support to its member countries. UNIDO established in 1983 an inter-agency Working Group on Biosafety (WGB) in collaboration with the United Nations Environment Programme (UNEP) and the World Health Organization (WHO) to which Food and Agricultural Organization later joined as a member. Other organizations have come up with biosafety guidelines. The Organization of Economic Cooperation and Development (OECD) published a document in 1986 on *Recombinant DNA Safety Considerations*. The European Commission (EC) released its directives in 1990 on the contained use and deliberate release into the environment of GMOs.

One of the aims of the WGB of UNIDO/UNEP/WHO/FAO was to promote environmental applications of GMOs for industrial development in a safe and ecologically sustainable manner. To this end WGB had taken several initiatives such as 1. release of a manual on *An International Approach to Biotechnology Safety*; 2. develop a voluntary code of conduct (VCC) for release of GMOs and its annotated version [1]; 3. organize annual training programmes on biosafety; 4. publish a biosafety guidebook on the release of GMOs; 5. help in setting up institutional (IBC) and national biosafety committees (NBC) in member countries; and 6. establish Biosafety Information Network and Advisory Services (BINAS).

The VCC outlined responsibilities of researchers, national authorities, and industry for safe application of GMOs and aimed at harmonization of biosafety guidelines. BINAS provide information support with databases on regulations and guidelines adopted in different countries, on national authorities responsible for taking decisions for use of GMOs, list of expert panels in risk assessment and management, field releases taking place around the world and their environmental impact, if any. BINAS also provide advice to IBCs, NBCs and industry to facilitate safe application of GMOs and facilitate consumer confidence and acceptance.

Arising from these initiatives in an expanded and improved form were the Anglo-Dutch proposals to support the development of voluntary guidelines for safety in biotechnology. These guidelines were aimed to assist governments to develop their own biosafety mechanisms in line with international cooperation and harmonization.

Agenda 21

The need for an international framework on biosafety was given due importance at the United Conference for Environment and Development [9] (UNCED), 1992. The Agenda 21 of the UNCED specifically endorsed the requirement for environmentally sound management of biotechnology. Chapter 16 of the Agenda 21 stated that there was need for development of internationally agreed principles on risk assessment and management of all aspects of biotechnology. Agenda 21 emphasized that when adequate and transparent safety procedures are in place, the public would be able to analyze the potential benefits and risks of biotechnology and derive maximum benefits from the technology.

Convention on Biological Diversity

Parallel initiatives on biosafety were taken by UNEP's Convention on Biological

Diversity [10] (CBD). The CBD, developed and ratified under the auspices of UNEP, was intended to promote conservation of biological diversity, sustainable use of its components and fair and equitable sharing of benefits arising on utilization of these resources using relevant technologies including biotechnology. CBD contained provisions for avoiding or minimizing possible risks from GMOs. Recognizing the potential for risks associated with biotechnology applications, the CBD outlined under articles 8g, 19.3 and 19.4 steps to be taken by member countries in matters relating to biosafety. These are to establish or maintain means to regulate, manage or control the risks associated with the use and release of living modified organisms (LMOs) which are likely to have adverse impact on health and environment. The CBD also suggested the need for a biosafety protocol, the modalities of which could include appropriate procedures such as advanced information agreement in safe transfer, handling and use of LMOs.

Cartagena Protocol on Biosafety

Accordingly, an internationally binding safety standards protocol entitled "Cartegena protocol on biosafety" (CPB) was adopted [8] by the CBD in the early 2000. The CPB is proposed to regulate the transboundary movement of specific categories of LMOs. LMOs intended for direct use as food, feed or for processing (LMO-FFP) are subjected to the 'Precautionary Principle'(PP) which allows action to be taken if there is a suspicion of any potential environmental damage even without having full scientific proof. Thus PP shifts the burden of proving the safety on to those who wish to release the GMOs. CPB enables setting up minimum standards for risk assessment and safety measures for the transboundary movement of GMOs. As per the CPB, the party that exports the GMO under the advanced Information Agreement (AIA) has to notify the importing party with information on the GMO such as identification methods, centers of origin or risk assessment reports. The importing party has the right to decide on the import after assessing the reports received on the GMO. Further, the LMO-FFP must be labeled appropriately including that these are not intended for release into the environment. CPB also proposed development of an internet-based Biosafety Clearing House (BCH), an information system on the GMOs and their movement through international trade. CPB further calls for resources from member countries for technical assistance and capacity building in adoption of biosafety measures.

Conclusions

The prospects for adoption of ecologically sustainable biosafety measures which are also safe to health and environment seem bright. Even the USA, which has long felt that GMOs do not generally pose untoward and insurmountable safety problems are reviewing their stand. A proposal to enact Genetically Engineered Food and Safety Act (GEFSA), which subject all transgenic components of bioengineered foods to premarket review as food additives, is gaining momentum. FDA of USA also states that if bioengineering produces a material, difference in the product, it should be labeled. Besides the USA, Canada and Argentina are the largest exporters of GMOs. The later two countries are signatories to the CBP. The UK and Europe and Japan are proceeding cautiously in approving genetically engineered foods. In the UK, an Advisory Committee on Releases to the Environment (ACRE) submitted a report on how to minimize risks of gene flow and reduce risks to biodiversity. Most developing countries are apprehensive about the release of GMOs although many of them including China, India and Thailand and few Latin American countries are actively testing them in the field. These countries are also investing in establishing research facilities that can produce GMOs within their countries. The developing countries are now more confident on biosafety with the adoption of CPB.

Problems in ensuring biosafety continue to remain at large. CPB does not cover other LMOs such as those used as pharmaceuticals. Biosafety evaluation still depends on the legal framework of individual countries. Many developing and least developed countries do not have national biosafety legislation or biosafety guidelines and necessary expertise in decision-making on proposals for introduction of GMOs for industrial development. The international organizations, NGOs and industrialized countries should consider providing resources and support in capacity building in these countries and instil confidence in them in using the frontier technologies in safe and ecologically sustainable manner for their economic development.

References

1. Subrahmanyam, D. (1992) Annotated Voluntary Code of Conduct, UNIDO.
2. Darmency, H. (1994) *Mol. Ecol.* 3: 337.
3. Mason, H.S. and C.J. Arntzen (1995) Trends in Biotech. 3: 388.
4. Levin, M.A. and H.S. Straus (1991) Risk assessment in genetic engineering, McGraw Hill.
5. Altieri, M. (1999) In Biotechnology and biosafety, World Bank Publication.
6. Borlaug, N. (1996) In Meeting the challenges of population, environment and resources: The costs of inaction, World Bank Publication.
7. Steinbrecher, R. (1996) The Ecologist, 26: 273.
8. The Cartegena Protocol on Biosafety (2000) Biotech. and Develop. Monitor, 43: 2.
9. United Nations Environment and Development (1992) *Research Report* No. 55.
10. United Nations Environment Programme (1992) Convention on Biological Diversity.

Radiobiology and Bio-Medical Research
Edited by K.P. Mishra
Copyright © 2004 Narosa Publishing House, New Delhi, India

17. Molecular Aspects of Alkaloid-DNA Interaction: Biological Significance

M. Maiti

Biophysical Chemistry Laboratory, Indian Institute of Chemical Biology, 4 Raja S.C. Mullick Road, Calcutta 700 032, India

Abstract: The molecular aspects of some biological active plant alkaloids with a right handed (B-form), left handed (Z-form) and left handed (HL-form) structures of DNA has been investigated by using various spectroscopic and thermodynamic measurements. Of these alkaloids, sanguinarine and aristololactam-β-D-glucoside bind strongly to B-form DNA structure with preference to GC sequence and do not bind to Z-form but they convert both the form back to the bound right handed form as evidenced from circular dichroic and absorption spectroscopy. Aristololactam-β-D-glucoside binds to H^L-DNA very strongly compared to B-DNA as revealed from spectroscopic and thermodynamic measurements. Binding isotherm obtained from spectroscopic data shows both the alkaloids bind to B-form polymer in a non-cooperative manner and the mode of binding found to be intercalative nature. These studies reveal that alternating GC sequence undergoes defined conformational changes and interact with the alkaloids which may be an important aspect in understanding their extensive biological activities

During recent years there have been great advances in elucidating the factors that govern the affinity and specificity of binding of many naturally occurring and synthetic compounds to DNA. One important class of these compounds comprises those that bind to DNA by a mechanism of intercalation. Such compounds are important tools in molecular biology and some are used for the treatment of cancer in man [31, 21]. Of the biologically active molecules, naturally occurring plant alkaloids occupy an important position because of their extensive medicinal value. Most of the alkaloids have promising antimicrobial, antitumour and a host of other biological activities. For example, Sanguinarine, a benzophenanthridine plant alkaloid has antitumor and antimicrobial properties [1, 33]. The relatively recently isolated aristololactam-β-D-glucoside (ADG) of the *aristolochia* group of alkaloids has attracted recent attention for their extensive biological activities [5, 6]. These two molecules are representatives of two different groups of alkaloids and have significantly different structure. Sanguinarine is a planner phenanthridine fused aromatic ring system, while aristololactam-β-D-glucoside has a sugar moiety and have the elucidation of their DNA binding characteristics, is important from the stand point of their biological activities.

It is now a clearly established fact that B-form can adopt a variety of unusual conformations depending on nucleotide sequence and environmental conditions [7]. Of the unusual structures so for known, perhaps, the most striking example is the Z-DNA, which is left-handed duplex conformation with Watson-Crick base pairs and can be generated in alternating purine-pyrimidine sequences under a variety of solution conditions [22]. X-ray crystal structure analyses in oligonucleotide crystals further confirmed the left handed Z-form structure [30]. Although a precise biological role for the Z-DNA has not yet been ambiguously identified [29], evidence for the existence of Z-DNA *in vivo* has been provided [11–13]. Both right-handed B-form and left-handed Z-form coexists in plasmids in *Escherichia coli* and they have also been detected in chromosomes from Drosophila and in metabolically active mammalian cells [9]. Protonation of DNA has been studied for several years [14–16] but has gained considerable significance from the observation of Chen [3] that a left-handed (Z-like) conformation could be generated in poly (dG-dC).poly(dG-dC) at acid pH. From the extensive UV spectroscopy and circular dichroic studies, author's laboratory has advanced a model of left-handed structure with the formation of Hoogsteen base pairing at low pH [10, 6] and itself is yet another polymorphic form and coined as H^L-form [6]. The model of H^L-form has recently been confirmed by the FTIR studies of Tajmir-Raiahi and co-workers [28] and by the Raman spectroscopic studies of Otto and co-workers [25]. Several authors [22–24] have also reported the biological relevance of the protonation induced conformational study of nucleic acids with respect to biological systems.

Using extensive UV-VIS spectroscopy, fluorescence spectroscopy, circular dichroic spectroscopy, thermal melting, viscometric and thermodynamic studies author's laboratory has demonstrated for the first time that both sanguinarine and ADG bind to B-form DNA by a mechanism of intercalation with a high preference to GC base pairs [25–33]. It has been shown that sanguinarine binds strongly to the B-form of DNA (Table I) and converts the Z-form and H^L-form back to the bind right-handed form ADG binds strongly to B-form structure (Table I) and converts Z-form to bind B-form, while it binds to H^L-form (Table I). Thus differential effect may be due to the presence of the flanking sugar moiety, the orientation of the chromophore in ADG at the intercalation site which is rather tilted, that is, imperfect [20, 23]. This difference in orientation of the alkaloid molecule at the intercalation site may be one of the possible reasons for the differential recognition of the different forms (Z- and H^L-form) of DNA by ADG.

Table 1. **Binding parameters of sanguinarine and ADG on the complexation with B-form and H^L-form DNA obtain from spectrophotometric studies**

Nucleic Acid	B-DNA-sanguinarine		B-DNA ADG		H^L-DNA ADG	
	$K'(10^5 M^{-1})$	n	$K'(10^5 M^{-1})$	n	$K'(10^5 M^{-1})$	n
poly(dG-dC)	14.5±0.8	3.2±0.1	4.5±0.4	4.6±0.3	6±2.0	3.0±0.1
poly(dG-m⁵dC)	35.0±3.0	2.6±0.1	35.0±4.0	4.0±0.1	15.0±2.0	3.0±0.1

Three determinations each K', the intrinsic binding constant and n, the number of nucleotide occluded was obtained according to excluded site model developed by McGhee and von Hippel (1974) for a nonlinear noncooperative ligand binding system using the following equation $r/C_f = K'(1 - nr)[(1 - nr)/(1 - (n - 1)r]^{(n-1)}$ where r and C_f are the bound drug and free drug concentration respectively.

The most important evidence for the intercalation model of binding is obtained from the viscosity measurements with linear sonicated H^L-DNA structure. Viscometric techniques are well established as a method to investigate the extension of DNA helix associated with intercalation [25]. The slope β is a parameter related to the functional increase in the contour length of rod-like linear H^L-DNA molecule induced by monofunctional intercalative ligands. The main result is that sanguinarine binds only to B-form DNA by intercalation, while ADG binds both B-form and H^L-form DNA in the same intercalative manner.

The temperature dependence of interaction has been used to derive the enthalpy of ADG binding to B-form and H^L-form of GC polymer and its methylated analogue. It has been observed that the binding of the alkaloid to B-form DNA is exothermic and the binding process is enthalpy driven (Table 2). The intermolecular interaction is characterized by a favourable negative enthalpy change and opposed by a negative entropy change in each B-form polymer ADG binding to both H^L-form DNA, is an exothermic process and the binding free energy arises primarily from a large negative enthalpy due to intermolecular interactions at the intercalation site. In both the H^L-form DNA, intercalation is favored by a negative enthalpic ($\Delta H°$) contribution and is opposed by the decrease in entropy ($\Delta S°$). Results of thermodynamic study also reveal that the H^1-form DNA structure is a highly ordered structure and ADG binds strongly to this structure compared to B-form structure (Table 2).

Table 2. **Thermodynamic parameters of sanguinarine and ADG on the complexation with B-form and H^L-form DNA obtain from spectrophotometric studies**

Nucleic Acid	$\Delta G°$(kcal/mol)	$\Delta H°$(kcal/mol)	$\Delta S°$(kcal/mol)
Sanguinarine B-DNA complexation			
poly(dG-dC)	−8.58±0.08	−14.31±0.23	−19.56±0.04
poly(dG-m⁵dC)	−8.38±0.08	−10.86±0.02	−8.83±0.1
ADG B-DNA complexation			
poly(dG-dC)	−7.76±0.02	−10.35±0.1	−8.70±0.04
poly(dG-m⁵dC	−8.98±0.05	−13.86±0.06	−16.38±0.05
ADG-H^L DNA complexation			
poly(dG-dC)	−7.3±0.05	−12.74±0.06	−19.23±0.05
poly(dG-m⁵dC	−7.70±0.05	−19.93±0.06	−32.62±0.05

Taken together, it is concluded that ADG differentially recognized the H^L-form DNA structures, whereas classical intercalator sanguinarine fails to do so. Thus ADG can be effectively utilized as a probe to recognize different kinds of structural polymorphs in DNA and if H^L-form structure plays a significant role in the control of the cell processes, ADG may convey some specified meaning for its regulatory role in biological systems.

References

1. Babich, H., H.L. Zuckerbraum, I.B. Barber, S.B. Babich and E. Borenfreund (1986) *Pharmacol. Toxicol.* 78: 397–405.
2. Cassady, J.M., W.B. Baird and C. Cheng (1990) *J. Nat. Prod. Lloydia*, 53, 23–41.
3. Chen, F.M. (1984) *Biochemistry* 23: 6159–6165.
4. Chen, Z. and D. Zhu (1987) *The Alkaloids* 31: 29–65.
5. Colins, J.S. (1992) *Agents Actions Spec.* C$_4$, 47.
6. Das, S., G.S. Kumar and M. Maiti (1999) *Biophys. Chem* 76: 199–218.
7. Foster, J.W. (1995) *CRC Rev. Microbiol.* 21: 215–224.
8. Herbert, A. and A. Rich (1996) *J. Biol. Chem.* 271: 11595–11603.
9. Jimenez-Ruiz, A., J.M., Requena, M.C. Lopez and C. Alonso (1991) *Proc. Natl. Acad. Sci. USA* 88: 31–37.
10. Kumar, G.S. and M. Maiti (1994) *J. Biomol. Struct. Dyn* 12: 183–201.
11. Li, L., D. Von Kessler, P.A. Beachy and K.S. Matthews (1996) *Biochemistry*, 35: 9832–9838.
12. Maiti, M., R. Nandi and K. Chaudhuri (1982) *FEBS Lett.* 142, 280–284.
13. Maiti, M., R. Nandi and K. Chaudhuri (1984) *Indian J. Biochem. Biophys.* 21, 158–165.
14. Maiti, M. and R. Nandi (1986) *Indian J. Biochem. Biophys.* 23: 322–325.
15. Maiti, M. and R. Nandi (1987) *Anal Biochem.* 164: 68–71.
16. Maiti, M. and R. Nandi (1987) *J. Biomol. Struct. Dyn.* 5: 159–175.
17. Marck, C., D. Thiele, C. Schneider and W. Guschlbauar (1978) *Nucleic Acids Res.* 5: 1979–1984.
18. Nandi, R. and M. Maiti (1985) *Biochem. Pharmacol* 34: 321–324.
19. Nandi, R., K. Chaudhuri and M. Maiti (1985) *Photochem. Photobiol.* 42, 497–503.
20. Nandi, R., S. Chakraborty and M. Maiti (1991) *Biochemistry*, 30, 3715–3720.
21. Neidle, S., L.H. Pearl and J.V. Skelly (1987) *Biochem. J.* 243: 1–13.
22. Pohl, F.M. and T.M. Jovin (1972) *J. Mol. Biol.* 67: 375–396.
23. Ray, A. and M. Maiti (1996) *Biochemistry*, 35: 7394–7402.
24. Saenger, W. (1984) *Principles of Nucleic Acid Structure*. New York: Springer-Verlag.
25. Segers-Nolten, G.M.J., N.M. Sijtsema and C. Otto (1997) *Biochemistry* 36: 13241–13247.
26. Sen, A. and M. Maiti (1994) *Biochem. Pharmacol.* 48: 2097–2102.
27. Sen, A., A. Ray and M. Maiti (1996) *Biophys. Chem.* 59: 155–170.
28. Tajmor-Riahi, H.A., J.F. Neault and M. Naoui (1995) *FEBS Lett* 370: 105–108.
29. Vander, J.H., G. Marel and A. Rich (1979) *Nature* 82: 680–686.
30. Wang, A.H.J., G.J. Quigley, F.J. Kolpak and J.L. Crawford, Van Boom,
31. Waring, M.J. (1981) *Annu. Rev. Biochem* 50: 159–192.
32. Wittig, B., T. Dorbic and A. Rich (1991) *Proc. Natl. Acad. Sci. USA* 88: 2259–2265.
33. Wolff, J. and L. Knipling (1993) *Biochemistry* 32: 13334–13339.

Radiobiology and Bio-Medical Research
Edited by K.P. Mishra

18. Fluorescent Labels in DNA/RNA Diagnostics

Krishna Misra

Nucleic Acids Research Laboratory, Department of Chemistry, University of Allahabad, Allahabad-211002, India

Latest advances in molecular biology and biotechnology are creating exciting possibilities for DNA/RNA diagnostics. Synthetic sequences of DNA/RNA are becoming extremely important because of their multipurpose applications not only in molecular biology or genetics but in therapy also. The study at DNA level analyses the genetic potential of individual humans and at RNA level it expresses information of particular cells. Nucleic acid diagnostics will certainly alter many aspects of modern medicine, including pre- and post-natal analysis of genetic diseases, the identification of individuals predisposed to conditions such as diabetes or coronary heart diseases and the analysis of infectious viral diseases ranging from common cold to cancer and AIDS.

Specific methods are available for monitoring Oligo sequences both *in vitro* and *in vivo,* inclusive of radioisotopic labeling and spectroscopic assays. The most commonly used method of labeling nucleic acids is by radioisotopes, primarily ^{32}P and ^{35}S, but due to a number of factors, for example, short half lives of isotopes, cumbersome nature of autoradiographic or scintillating counter methods of detection along with health hazards and disposal problems, limits the use of this technique despite its' high sensitivity.

Although a large number of reporter groups are now available for nucleic acid probes like spin labels and triplet labels, yet fluorophores are most attractive because of their ease of application, stability, direct detection and discriminable emission spectra, despite their relatively lower but comparable levels of sensitivity. There are certain advantages of fluorescence based sequencing over radioactive based sequencing. Fluorescence provides sufficient sensitivity for real time optical detection of the small amount of DNA present in DNA sequencing gels ($\sim 10^{-15}$ mole per band). Four different fluorophores are used for four different bases and identification done by colour (Rhodes machine) from which direct inference can be drawn. Fluorescence measurements can give information regarding conformations, binding sites, solvent interactions, degree of flexibility, intermolecular distances and the rotational diffusion coefficient of macromolecules. With living cells fluorescence can be used to localize otherwise undetectable substances.

Several factors determine the quantum yield (Q) of fluorescence, some of these are the properties of the molecule itself (internal factors) and some are environmental. The nucleic acid molecules being quite flexible have very high vibrational levels of the ground state and, therefore, the excitation energy is dissipated resulting in no fluorescence. However, biochemists are more concerned with environmental factors of these macromolecules. The effect of environment is primarily to provide radiation-less processes that compete with fluorescence and thereby reduce Q, that is, effect quenching. In biological systems quenching is usually due to collision with exchange of energy or a long range radiative process called FRET (fluorescence resonance energy transfer). Thus, effect of solvent or dissolved compounds (called quenchers), temperature, pH, neighbouring chemical groups or the concentration of the fluorophore are important environmental effects in the study of macromolecules.

Two types of fluors are used in fluorescence analysis of DNA/RNA's:

(i) intrinsic fluors: contained in the macromolecules themselves,
(ii) extrinsic fluors: added to the system, usually binding to one of the components.

Intrinsic fluors

Some nucleosides act as self fluorescent moities, for example, 2'-deoxyinosine, 2'-deoxyisoinosine, 1, N-etheno-2-aza-adenosine, some fused pyrimidine compounds, 1-N-guano-2-azaadenosine and 1, N -ethenoadenine. All these nucleosides are highly fluorescent. 2'-Deoxyinosine has been effectively used in solid phase synthesis. A pyrido-pyrimidine deoxynucleoside, namely, 3-D-2'-deoxyribofuranosyl-2,7-dioxopyridol [2,3-d] pyrimidine and its ribo counterpart are fluorescent. This pyrido-pyrimidine analogue is considered to form hydrogen bonds with guanine and adenine and shows absorption at 340 nm.

Extrinsic fluors

In many cases a fluor is introduced in a DNA/RNA molecule either by chemical coupling (through linkers) or by simple binding (intercalators/groove binders). Certain basic requirements for such fluors must be met, for example, (i) these fluors must be tightly but reversibly bound at a unique location through linkers, (ii) its' fluorescence must be sensitive to environmental conditions and should not quench on covalent binding, and (iii) it should not itself affect the features of the macromolecule being investigated and must not interfere with the process for which it is used as diagnostic tool, for example, hybridisation, duplex/triplex stabilities and electrophoretic behavior. These criteria must be verified. Such fluorescent probes generally used are planar molecules having polycyclic aromatic systems of about the size of a base pair. These are generally water soluble, chemically and thermally stable molecules. Some examples of fluors commonly used for labeling nucleic acids are fluorescein, biotin, rhodamine, ethidium bromide, acridines, ANS(1-anilino-8-naphthalene sulphonate), DNS (dansyl chloride), Hoechst and TNS (2-p-toluidyl-naphthalene-6-sulphonate).

We have used a large number of intrinsic as well as extrinsic fluors in our laboratory for various studies. A large number of naphthalimide derivatives have been synthesized, characterized and used for labeling oligos. Following are some of these:

X = NO$_2$ (I) / NH$_2$ (II), published in
Nucleosides Nucleotides 10, 963, 1994

R = –C$_6$H$_4$–COOH (III)/–(CH$_2$)$_5$–COOH
(IV) published in *Neurochem Int.* 31/3,
405, 1997

R = p-Toluenesulphonyl (V), Dansyl (VI) and Dabsyl (VII), unpublished result

There are various applications of nucleic acid labeling, which fall into three main categories,

1. To monitor enzymatic reactions (cDNA synthesis),
2. For structural and functional analysis of genes, and
3. As hybridization probes to locate the target sequence.

A simple example of the use of acridine orange as fluor is for determining strandedness of polynucleotides. When bound to polynucleotides acridine orange shows an increase in Q and a shift in λ_{max}. When saturating amounts of acridine orange are added , the values of λ_{max} are significantly different for double (green fluorescence) and single (red fluorescence) stranded polynucleotides. If a sample contains both, double and single stranded polynucleotides, the fluorescence spectrum will have two peaks, one for each value of λ_{max}. Thus, if a high concentration of acridine orange is added to a cell, the nuclear material fluoresces green whereas the RNA containing cytoplasm appears orange (Fluorescence Microscopy). This has been used with eukaryotes to observe nucleic acid and chromosomes and to detect RNA in the nucleus; it has been used in prokaryotes to localize DNA.

The fluor known as SITS has been used for distinguishing living from dead cells. SITS is taken up by living cells but is restricted to small vesicles, which thereby fluoresce. If the cell is not living, the vesicles are usually disrupted, as a result the SITS diffuses freely through the cell. It binds tightly to nuclear membrane and fluoresces brightly. Thus, a fluorescent nuclear membrane is criterion for cell death.

Ethidium bromide binds tightly to DNA and thereby increases quantum yield. This increase is linear throughout a wide range and thus DNA quatitation can be effected. The enhancement of Q is also used to detect DNA on PAGE or Agarose by immersing gels into ethidium bromide solution and subsequently exposing it to exciting light.

If the sample is excited by polarized light, the resultant fluorescence is only partially polarized or completely unpolarized. The magnitude of the change of polarization gives information about the physical state of matter. Polarization (P) is always less than 1, that is, P<1 and this is called fluorescence depolarization. With increasing mobility polarization decreases. Orientation of DNA in chromosomes can be determined by polarization of extrinsic fluorescence.

Under certain circumstances energy absorbed by one molecule (a donor) can be transferred to another fluor (an acceptor) at some specific distance away. This phenomenon called resonance energy transfer (RET), a necessary but not sufficient condition of which is that the emission spectrum of donor overlaps the absorption spectrum of the acceptor. The efficiency of transfer is a function of the separation of fluors and, therefore, is used for measuring molecular distances. The efficiency of transfer, $\xi = R_0^6 / (R_0^6 + R^6)$

Where R is distance between donor and acceptor and R is a constant related to each donor-acceptor pair that can be calculated from certain parameters of the absorption and emission spectra of each. Therefore, from equation $R = R_0 [1 - \xi/\xi]^{1/6}$ the molecular distance between donor and acceptor can be calculated. Combination of laser beams with FRET is likely to make possible the measurement of distances in the range 5–9 nm. Inevitably, fluorescence has the potential to be the method of choice for diagnostics not only in nucleic acids but also in the solution of all complex biochemical processes.

Radiobiology and Bio-Medical Research
Edited by K.P. Mishra

19. Fluorescent Chimeras and Living Colors: Unraveling the Mysteries of Cell Signaling (The Story of Phospholipase C-δ1)

Srinivas Pentyala*, Edward Tall, Shobha Mathew, Kavita Tanguturi, Praveen Yalamanchili, and Mario Rebecchi

Department of Anesthesiology, Physiology & Biophysics School of Medicine, State University of New York Stony Brook, NY, USA

Abstract: Using conventional biochemical techniques to understand the role of signaling molecules in cell biology has long been considered the hallmark of basic science research. Over the years, these time-tested methods allowed us to understand in detail, the functional role of several of the biological molecules. However when the same concepts and techniques were applied to understand the regulation of molecules in living cells, researchers often find that cellular regulation of a molecule does not always correlate with the hypothesized functions as predicted by the virtue of its prior biochemical characterization. Still, biochemical characterization in conjunction with molecular biology is considered a powerful tool to understand the mechanism of several of the signaling molecules. Recent development of Green Fluorescent Protein (GFP) technology and its versatile applications has revolutionized the fields of biotechnology and cell biology. Combining the existing tools of Immunocytochemistry and Microscopy in addition to the well-established techniques of Biophysics, Biochemistry and Molecular Biology, one can now use the fluorescent GFP chimeras to visualize their living colors to unravel and understand the mystery of cellular signal transduction. Functional autopsy of a key signaling enzyme, Phospholipase C-δ1, using both old and new techniques is presented here as an example correlating our understanding of this molecule with progressive application of techniques culminating with GFP technology.

Introduction

Green fluorescent Protein (GFP), cloned from the jellyfish *Aequorea victoria* is now considered a versatile and a powerful tool of biotechnology in studying the expression of various proteins. GFP is easily visualized when excited with UV light and its fluorescence does not depend on any exogenous compounds. This protein, when expressed alone, appears cytosolic, with no cellular targeting mechanisms of its own, making it an ideal tag for a protein of interest. The brilliant, intrinsic fluorescence of GFP has made it useful for molecular and cell biologists as a non-invasive indicator of gene expression. Enhanced GFP variants are available in green (EGFP), yellow (EYFP), and cyan (ECFP), which are stable, up to 30 times brighter than wild-type GFP, have longer wavelengths and can be used for fluorescence resonance energy transfer. Unlike other bioluminescent reporters, these proteins do not require additional proteins, substrates, or cofactors to emit a signal. They are ideal for monitoring gene expression and protein localization *in vivo*, *in situ*, and in real time. Additionally, GFP is an excellent reporter because its fluorescence can be detected directly, is species independent, and is stable over time. Since its first introduction in molecular biology, GFP has been expressed in many organisms, including bacteria, yeast, slime mold, many plants, fruit flies, zebra fish, many mammalian cells, and even viruses. Moreover, many organelles, including the nucleus, mitochondria, plasma membrane, and cytoskeleton, have been marked with GFP. GFP has initiated a revolution in molecular cell biology by establishing a general approach for visualizing nearly any protein of interest in any cell, tissue or species. Researchers working at all levels of biology, such as single molecule dynamics, protein trafficking within cells, organelle dynamics and cell and tissue behaviours during development, have made important use of GFP (Aizawa et al. 1997; [1, 3, 6, 15, 26, 27, 28, 29, 33, 36, 35] Feng et al. 1998; Oancea & Meyer 1998; Servant et al. 1999; Shen et al. 1998; Stauffer et al. 1998; Tall et al. 2000a; Venkateswarlu et al. 1998; Wang et al. 1997; Wang et al. 1996). Tagging GFP to molecules involved in signal transduction has revolutionized the field of cell biology, opening new horizons for researchers in the area of cellular signal transduction, where it is now possible to actually chase their molecule of interest in live cells in real time.

Phosphoinositide-specific phospholipase C (PLC) isozymes found in eukaryotes are a group of key signal transduction enzymes. They cleave the polar head group from inositol phospholipids. Under the control of cell surface receptors, these enzymes hydrolyze the highly phosphorylated lipid, phosphatidylinositol 4, 5 bisphosphate (PIP_2), generating two intracellular products, inositol(1, 4, 5)trisphosphate (IP_3), a universal calcium-mobilizing second messenger and diacylglycerol (DAG), an activator of protein kinase C. In the late 80's and early 90's, several mammalian PLC subtypes were isolated and their corresponding cDNA sequences determined. PLC subtypes are found in eukaryotic organisms ranging from yeast and slime molds to higher plants and mammals. Classified as β, γ or δ they are soluble multi-domain proteins ranging in MW from 85 kDa to 150 kDa. Four β, two γ, and four δ isoforms, as well as numerous spliced variants, have been described in mammals. Those found in yeasts, slime molds, filamentous fungi, and plants closely resemble mammalian δ.

The eukaryotic PLCs have a modular domain organization. With the exception of a few spliced variants, all eukaryotic PLCs contain an amino terminal pleckstrin homology (PH) domain followed by four EF hand motifs, an X/Y motif constituting a catalytic/barrel, and a single C-2 domain (Scheme 1). Structures of the PH domain of PLC-1 and a truncated form lacking this domain have been separately solved (4). Comparisons of their DNA sequences suggest an evolutionary relationship in which the δ subtype appeared first in primitive single-celled eukaryotes. At present, many of the players in phosphoinositide/calcium signaling are identified, some with three-dimensional pictures. On a cellular level, questions

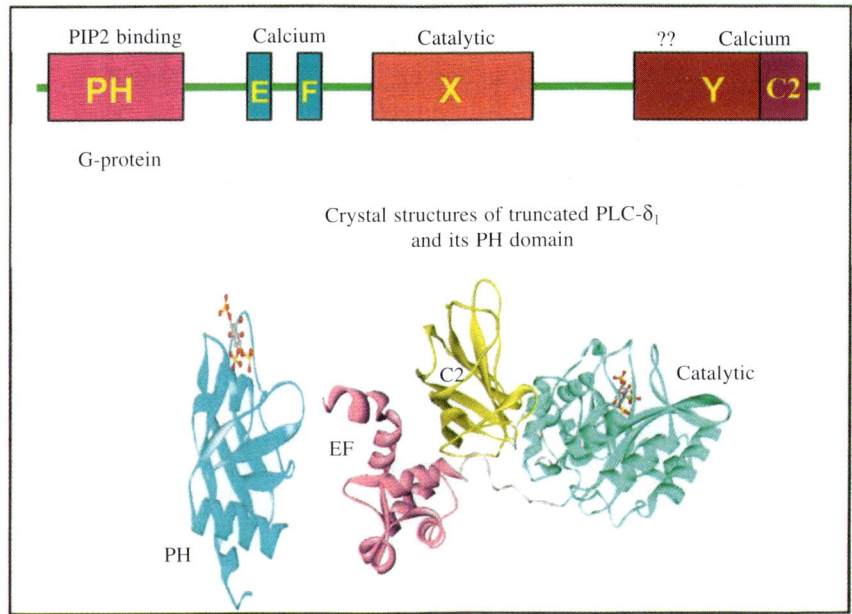

PIP2 binding   Calcium   Catalytic   ??   Calcium

PH   E F   X   Y   C2

G-protein

Crystal structures of truncated PLC-δ_1
and its PH domain

C2   Catalytic

EF

PH

Scheme 1.   Structure/Function of PLC-δ1

of which PLC isozymes go with which regulators are mostly answered (see our recent review on PLC, [23]. Despite this progress, our understanding of how and where PLC isozymes work in living cells is limited.

The PLC δ isozymes range in size from 64 to 101 kDa, and are found in all eukaryotes from yeast to humans. While a single δ-related gene is found in yeast, cellular slime molds, hydra, filamentous fungi, various mammalian species, and plants express numerous isoforms, including alternatively spliced variants. Although much is known of its structure and chemistry, particularly the mammalian δ_1 isoform, the biological role has remained obscure. Thus far, most clues to function have been discovered in non-mammalian organisms.

Through the work of many different investigators, we have learned how PLC isoforms act as catalysts, discovered what proteins and lipids regulate their activities, and gleaned some hints of their diverse biological roles [5, 7, 8, 11, 13, 16, 19, 30, 32]. Crystallographic studies of PLC-δ_1 catalytic core and its constituent domains have offered us a molecular view of the reaction and provided a template for interpreting the structure and function of similar modules in the other subtypes. The current challenge is to understand the nature and dynamics of the membrane/enzyme microinterface, and their relation to the cycle of substrate binding and product release. While enormous progress has been made in uncovering the how, what, and where of the PLCs, many questions remain unanswered. Most of these concern their operation in living cells. What are the many PLC δ isoforms and their variants doing in higher plants and animals? Are they signal amplifiers? Do they participate in calcium oscillations? New information suggests a higher level of organization than is implied by the current regulatory schemes, giving rise to a number of questions: Are these freely diffusing effector proteins or they are part of a highly organized network? Do these enzymes only act at the plasma membrane? Do they act in concert? Where is their substrate localized and how is it supplied? Finally, what is the physiologic function of each of these many isotypes and their individual domains and how is their tissue specific-expression controlled?

In order to answer same of the above mentioned queries and to understand the critical role PLC-δ1 isozyme plays in cellular physiology, a systematic approach was taken initially to biochemically characterize this molecule. Over the years, many investigators (including us) reported the *in vitro* regulation of the whole enzyme in terms of its ability to hydrolyze lipids [20, 21, 22, 23, 24, 25, 29]. Recently we reported the functional aspects of individual domains of PLC-δ1 in terms of its binding interactions with lipids and other signaling molecules [36, 37, 38]. However majority of these studies were limited to *in vitro* characterization of this enzyme. Recently, using GFP technology we started studying the physiological regulation of this enzyme in living cells in terms of its localization, translocation, role of individual domains and interaction with other proteins [30]. Here, we present the approach we had taken and some of the observations that were made with regard to the physiological role of PLC-δ1 and its individual domains in living cells using GFP technology.

Methods

Construction of prokaryotic and eukaryotic expression plasmids and purification of recombinant proteins: A collection of mammalian expression plasmids encoding native as well as fusion proteins of EGFP were made by cutting and splicing the gene for recombinant human PLC-δ_1 or Actin and ligating into pGFP-N1 mammalian expression vector under the control of a CMV promoter, which was human codon optimized. The pGFP-N1 vector from Clonetech (4.7 kb) confers kanamycin resistance. Plasmid vectors were used to transform DH5alpha-FT competent cells grown in 3.5 ug/mL Kanamycin. DNA was harvested from 0.5 or 1.0 L cultures by either CsCl centrifugation method or by using Qiagen DNA purification kits. PH domain lacking δ_1 was made by deleting PH domain. The constructs were introduced into cells by electroporation and checked for transient transfection by fluorescent microscopy, western analysis and immunocytochemistry [30]. Corresponding bacterial expression plasmids were also constructed to validate properties and the functions of the individual domains of the different PLC isozymes (β_1, β_2 and δ_1) using affinity tags like a His tag or a GST tag, and fluorescent GFP tags. Recombinant PLC-δ_1 and fusion proteins are prepared as described earlier [37, 38] Wang (1999b) and proper folding of the protein products was confirmed by circular dichroism using the apparatus at the National Synchrotron Light Source at Brookhaven National Laboratories (Upton, NY) on an Aviv 62A DS spectrometer (Scheme 2).

Cell Culture and Electroporation: Different cells (NIH-3T3, PC12 and NIE115 cells) are cultured to 80–90 percent confluence using a DMEM enriched medium

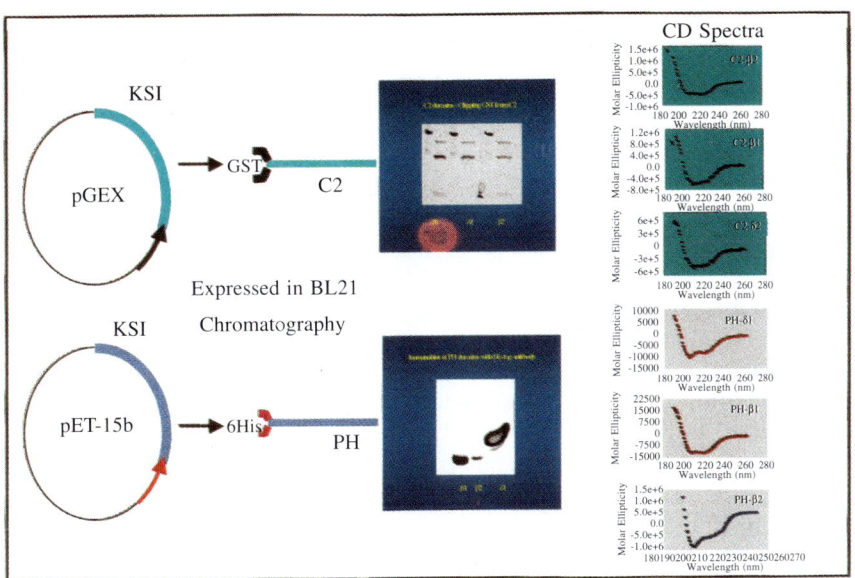

Scheme 2. Expression of recombinant proteins (GST-C2 & 6His-PH Domains)

(10 percent fetal bovine serum, 2 percent horse serum, 1mM sodium pyruvate, 1 mM non-essential amino acids, 100 units/mL penicillin, and 100 μg/mL streptomyosin) at 37°C and 10 percent CO_2. Just prior to electroporation, cells are washed once in intracellular electroporation buffer (ICEB): 125 mM KOH, 4 mM NaOH, 73 mM PIPES, 34 mM myoinositol, 10 mM $NaHCO_3$, 5 mM K_2HPO_3, 5 mM KH_2PO_3, 5 mMD-glucose, 4 mM $MgCl_2$, 1 mM $MgSO_4$, 1 mM $CaCl_2$, pH 7.0. These are then distributed evenly into several conicals, yielding –6.25 cm^2 cells per transfection cuvette, centrifuged once again, and resuspended in 0.5 mL ICEB ± plasmid DNA [75 ng/mL]. The 0.5 mL cell suspensions are transferred into 0.4 cm cuvettes (BioRad, CA) and electroporated by a BioRad Gene Pulser set at 500V/cm and 960 μF for 1–2 pulses, 30 seconds apart, at room temperature. Cells are then washed and resuspended in normal growth medium with or without NGF. Cells are plated into LabTek chambered coverglasses (NUNC, IL), coated with human fibronectin (Boehringer Mannheim, IN). The following day, cells are washed twice with D-PBS (Gibco) and refed. are used in our experiments.

Fluorescence activated cell sorting (FACS): To demonstrate the utility of GFP in flow cytometry, we transfected near-confluent cells with either vectors containing wild type GFP and PLC-GFP, respectively. 48 hours after transfection cells were lifted from their culture conditions by trypsinization, washed with PBS and fixed with 1 percent formaldehyde. We compared the profile of mock-transfected cells to that of cells transfected with GFP constructs using a FACSCAN analyzer (Becton Dickenson).

Subcellular Fractionation: For the separation of cytosol, nuclei and membranes, monolayers are detached and cells homogenized in pH 7.4 buffer (Tris 20 mM, NaCl 200mM, 0.3 M sucrose, 5 mM DTT, 1 mM MgCl2, 1 mM EGTA, 1x mammalian cell protease inhibitor cocktail (SIGMA), and 2 mM PMSF) by 10 passages in a tight fitting Dounce. The concentration of sucrose is brought to 1.8 M and the sample then layered over a 2 M cushion in a centrifuge tube and covered with 1.6 M sucrose. The tube is then filled with 0.3 M sucrose. After centrifugation for 2 h at 35,000 rpm in an SW41 rotor, nuclei are recovered in the pellet, the cytosol in the loading cushion and microsomal membranes in the 1.6 M cushion, up to its interface with the 0.3 M sucrose. The results are normalized to total protein and compared.

SDS-PAGE and Quantitative immunoblotting: Cells and subcellular fractions are subjected to SDS-PAGE and electrotransferred to PVDF membranes. The membranes are blocked in 5 percent dry milk in TBS and probed with various anti PLC-δ_1 antibodies. Our two best antibodies are a mouse Mab that recognizes the C_2 domain, and a sequence-specific rabbit polyclonal antibody, which recognizes the PH domain with high affinity and specificity [2]. The blots are probed with goat anti-mouse or rabbit IgG, labeled with HRP. The blot is then developed in ECL reagent (Amersham) and scanned/quantified using a Kodak Imaging Station Model 440 CF as described earlier [18]. Luminescence intensities are compared to the signals from known amounts of PLC.

Epifluorescence microscopy: We used a 1520 x 1080, 12 bit CCD camera, Olympix AstroCam (LSR, UK) supported by Esprit imaging software, which is capable of 50–100 ms frame capture/transfer rates. This camera is used for high-resolution work. Frames are stored on a high- end PC, for processing with commercial software: Esprit, Autodeblur Silver, Autoquant, and Autovisualize 3-D. Morphometric analysis of exported TIFF files is performed using the NIH image package. Point spread functions (PSF) are obtained for each lens/bandpass combination using appropriately labeled latex beads, typically < 0.2 um diameter. Background images are taken before and after the experiment. The PSF and background images are critical to removing out-of-focus light and enhancing sensitivity (described below). For high-resolution imaging, 1520×1080 frames are processed as 12 bit images. After background subtraction, a computational confocal technique was used to remove out-of-focus light. The Fourier transforms of the PSF functions, known as the contrast transfer functions (CTF) was used in a nearest neighbor approach [70]. Fourier domain transforms of the two neighboring Z axis sections are convolved with their corresponding out-of-focus normalized CTFs. When needed, windowing of the images will be performed to prevent edge artifacts. The resulting images are subtracted in the Fourier domain from the in-focus intensity information to obtain the final deblurred image. Corrections in the CTF may be needed for phase shifts, since the image must remain phase invariant. Sharpening of the deblurred image was performed using the in-focus PSF in the Fourier domain. The final image is obtained by reverse transforming the CTF convolved image under non-negativity constraints. To resolve dynamic cell processes under time constraints that limit the number of images, a faster, though less accurate, so-called no-neighbors approach can be used with a single out-of-focus PSF operating on one in-focus image.

Indirect Immunofluorescence: Cells are fixed the same day immediately after live imaging. Cells are washed once in 37°C Optimem medium without any supplements and fixed for 20 minutes in 4 percent formaldehyde (prepared from paraformaldehyde) in Ca^{2+}-free, Mg^{2+}-free PBS. They are washed 3 times for a total of 5 minutes, and permeabilized with 0.1 percent TX-100 for 5 minutes and washed 3 times again with PBS and blocked with 1 percent goat serum. Following incubation with primary antibodies, the specimens are incubated with Texas-Red goat anti-mouse or rabbit IgG (Molecular Probes) in PBS with 0.5 percent BSA for 1 hour, followed by 3 washes with PBS for a total of 20 minutes and mounted with VectaShield medium (Vector Laboratories, CA).

Confocal Microscopy: Cells were prepared as described above for epifluorescence microscopy or fixed as described for immunocytochemistry. Imaging was conducted on a Nikon Diaphot inverted microscope using Plan Apo 60, 1.4 N.A. oil objective lens (Nikon) at University Microscopy Imaging Center facility (UMIC, Stony Brook, NY). Excitation light at 488 nm and 529 nm was produced by an Odyssey confocal laser source (Noran Instruments, WI). Samples were scanned along the z-axis in 0.2 μm steps from top to bottom of the cell. Emitted fluorescence passed through either a 525 DF30 bandpass or a 600 EFLP cut-on filter (Omega Optical Inc., VT). Select images were imported into Photoshop for processing and display. Construction of a 2-dimensional image from confocal microscopy data was accomplished by using image analysis software by assigning color codes to fluorescence intensity. Furthermore, the image can be adjusted (by rotation, inversion) to show the fluorescence from different angles, and even have some slices removed to show the interior of the cell. An example of this was published by us in which we are looking to see within a living mammalian cell where a pool of a particular molecule is located [38]. Since the cells are living while under examination, the investigators can also view where and when the molecules' position changes in real time.

Results and Discussion

Cell fractionation indicates a stable pool of membrane bound enzyme: PC12 cells grown to 80 percent confluency in a T-75 flask were harvested for cell fractionation using the standard protocol of trypsinization. The cells were then washed with PBS and suspended in 500 ul of homogenizing buffer and homogenized by douncing. The homogenate is separated into two aliquots and to one fraction, 20 mM IP_3 is added and both the aliquots are processed for different soluble and insoluble fractions. Initial separation is done at 15,000 RPM for 5 minutes and the resultant sup is spun at 100,000 RPM for 1 hour to get a high-speed sup and pellet. Equal amounts of protein are subjected to SDS-PAGE and immunoblotted with δ_1 antibody. Immunoblots reveal a complex pattern of distribution of δ_1. Both the low speed and high-speed insoluble fractions revealed considerable amounts of PLC being associated with them and IP$_3$ treated fractions show that δ_1 associated with high speed membrane fractions can be extracted into the supernatant (Fig. 1A).

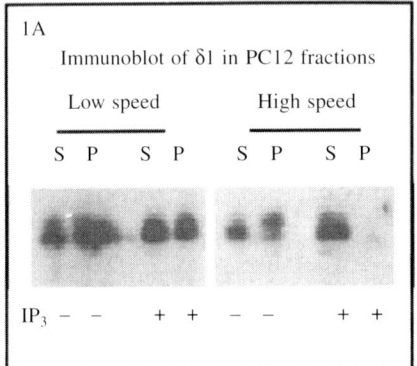

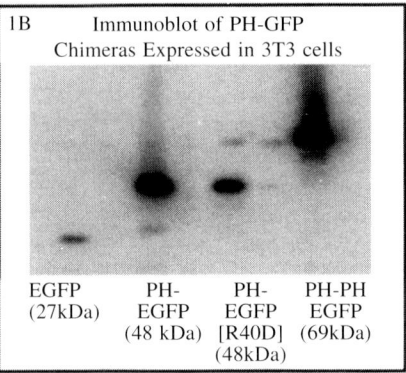

Figure 1. Localization, IP3 affinity and intactness of PLC and fusion constructs in PC 12 cells

Expressed Chimera constructs are intact in mammalian cells: Different GFP tagged chimera constructs that were introduced into mammalian cells (Fig. 1B) were checked for their expression by harvesting the cultures and subjecting the extracted protein to SDS-PAGE and Western blotting. Our results indicate that the fusion proteins, which are transiently expressed, are stable.

PLC-δ_1 and its PH domain follow a complex distribution pattern by GFP and Immunofluorescence: Expression plasmids were transfected into NIH-3T3 and PC12 cells by electroporation. The transfection efficiency was found to be in the range of 15 to 30 percent for fluorescent chimeric proteins (Fig. 2). Cells growing in chambered glass cover slips were viewed by epifluorescence microscopy 24 hrs post transfection. A Superfluor 40 X 1.3 NA lens was used. Cells expressing PLC-δ_1-GFP in 3T3 cells and PH-GFP in PC12 cells showed prominent localization to membrane ruffles, and non-uniform cell border localization and often punctate (Figs. 3A and 3B). LSCM confirms that many of these structures are vertical extensions of the plasma membrane (not shown). Growth cones containing substantial levels of PIP2 are visible in differentiated PC12 cell (Fig. 3B). Cells expressing dsRed-Actin (Fig. 3C) show both cortical and cable-like actin structures. To further characterize the subcellular distribution of PLC-δ_1, cells were fixed, permeabilized, and probed with PLC-δ_1 Mab and Texas-red labeled anti-mouse IgG (Figs. 4A and 4B). Note the punctate pattern across the membrane surface that is typically observed in addition to actin-supported structures. Basolateral membranes from PC12 cells anchored to poly-lysine coated glass were also prepared [9]. Their cytoplasmic surfaces reveal a similar punctate pattern (Fig. 4C). As control, the levels of fluorescence intensities could be manipulated by heterologous transfection with PLC-δ_1 expression plasmids. Fluorescence was undetectable in trivial controls, such as second antibody alone.

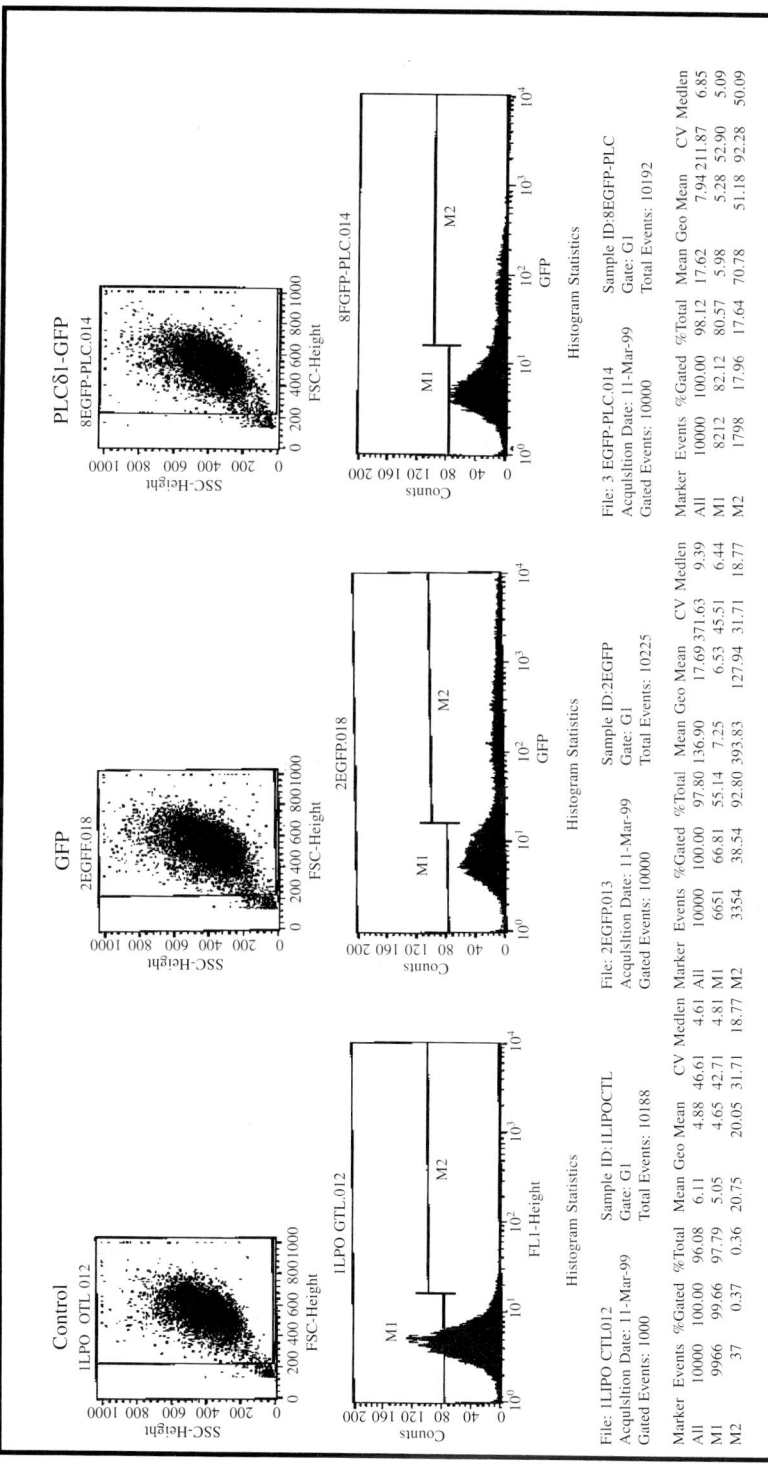

Figure 2. Transfection efficiency determination by FACS analysis

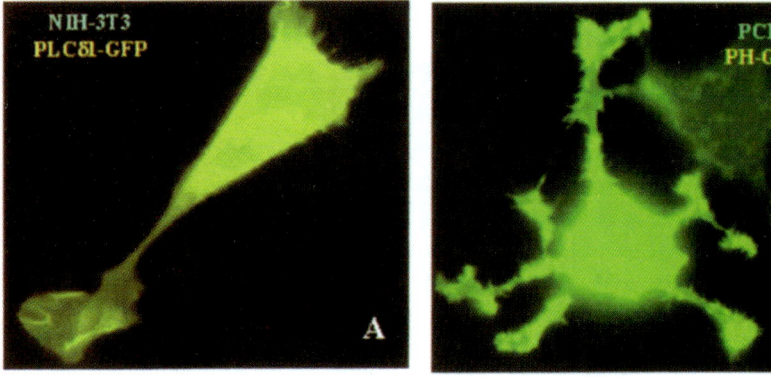

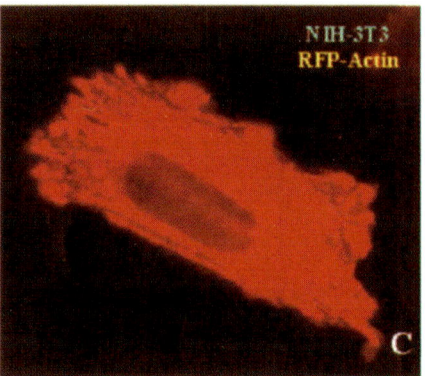

Figure 3. Live Cells—Fluorescence Imaging

PH-EGFP co-localizes with actin in PC12 cells: Cells transfected with PH-EGFP were fixed and probed with Alexa 594-Phalloidin, which binds to F-actin. Viewed with our GFP/TxRed filter set a prominent colocalization (yellow) of PH-GFP (green) with F-actin (red) is observed (Fig. 5). The results are similar to those obtained with NIH-3T3 fibroblasts [38].

Dual imaging of PI(4,5)P$_2$ and actin: We obtained a series of observations of living cells expressing both the δ_1PH-GFP and dsRed-Actin. While a few images were acceptable, the efficiency of expression of the dsRed actin was disappointing (the experience of other investigators as well). We have since employed a YFP-actin, based on the enhanced GFP, with additional mutations to red shift the absorption and emission spectra. This probe is paired with CFP linked to the δ_1PH domain. Dual band pass filters are used to separate the colors. The results are very promising. Both the cyan labeled PH domain and the yellow actin colocalize to membrane ruffles and other actin supported structures (Fig. 6). The combination of cyan and yellow constructs is also ideally suited for FRET studies.

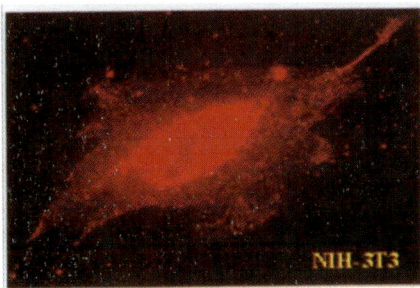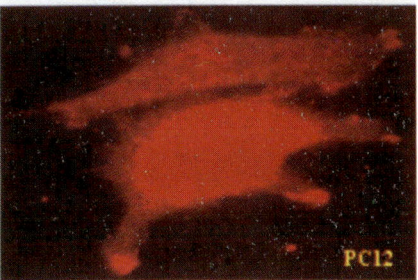

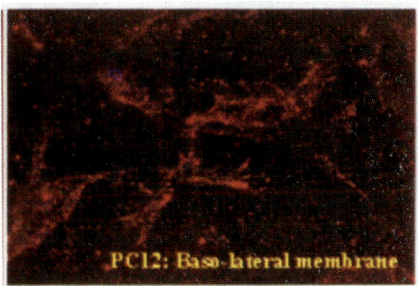

Figure 4. **Immunocytochemistry with PLCδ1-antibody showing pattern of distribution**

Gradient distribution of PIP$_2$ as measured by using PH-GFP construct: Given that the PH domain binds with high affinity to PIP$_2$ [12, 14, 23, 29–30], one can follow the location and movement of PIP$_2$ within a living cell by observing the location and movement of the green-glowing PH-GFP within that cell. In order to see where PIP$_2$ is within a living cell, we monitored these molecules using a PH domain of PLC-δ_1 (which binds with high affinity to PIP$_2$) that has a fluorescent GFP molecule attached to it. Using the image analysis software to give colors to pixel intensity we were able to localize PH-GFP, there by PIP2, in different areas of the cell (Fig. 7). This fluorescence is usually not uniform throughout the cell, but concentrated in some areas and missing from other areas. This difference in fluorescence intensity refers to the concentration gradient of GFP-PH domain there by PIP2. Using this construct we were able to monitor the levels of PIP2 in living cells.

Conclusion

Using PLC-δ_1 as model molecule, we tried to systematically characterize the functional as well as physiological role of this enzyme in a living cell. Over the years, using conventional biochemical, biophysical and molecular biological techniques, much of the characterization was performed on this molecule which enabled us to design studies to understand the physiological role this protein plays in live cells. Employing GFP technology to our studies made it possible to look for functional regulation of this key signaling molecules in live cells. The

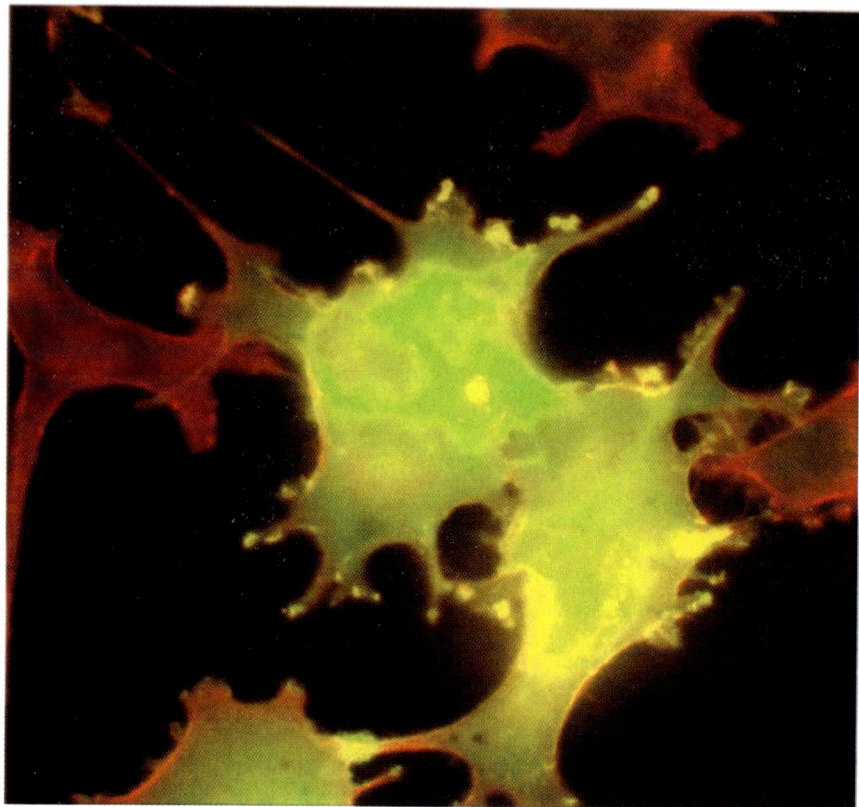

Figure 5. Colocatization of PH-GFP with Actin

high affinity of one of the domains of this molecule to bind to a membrane integrated lipid PIP2, allowed us to monitor the localization and translocation of the enzyme as well as its substrate, PIP2. We are also able to find sites of translocation and localization of this protein, which will enable us to look for potential binding/regulatory partners of PLC-δ_1.

Signal transduction proteins have been tracked to discover how cells sense spatial gradients. One of the incredible strengths of our model system is the capacity to track the dynamic behaviors of signaling molecules in living cells. The capacity for tracking living cells and phenotypic complementation of null mutants with GFP fusion proteins provides extremely useful tool. The optical clarity displayed by the cells facilitates real-time three-dimensional imaging. Also, all the GFP tagged protein expressing cells are ideal for use in flow cytometry. The 488 nm argon laser used in many standard flow cytometry machines efficiently excites DsRed, EGFP, and EYFP. EGFP and DsRed are the recommended combination to use when sorting a mixed cell population. Although the same argon laser excites EGFP and DsRed, the two proteins can be detected separately in the conventional FL1 and FL2 channels. Fluorescent protein tags offer several options for multiple labeling. For dual labeling, EGFP and DsRed—the brightest

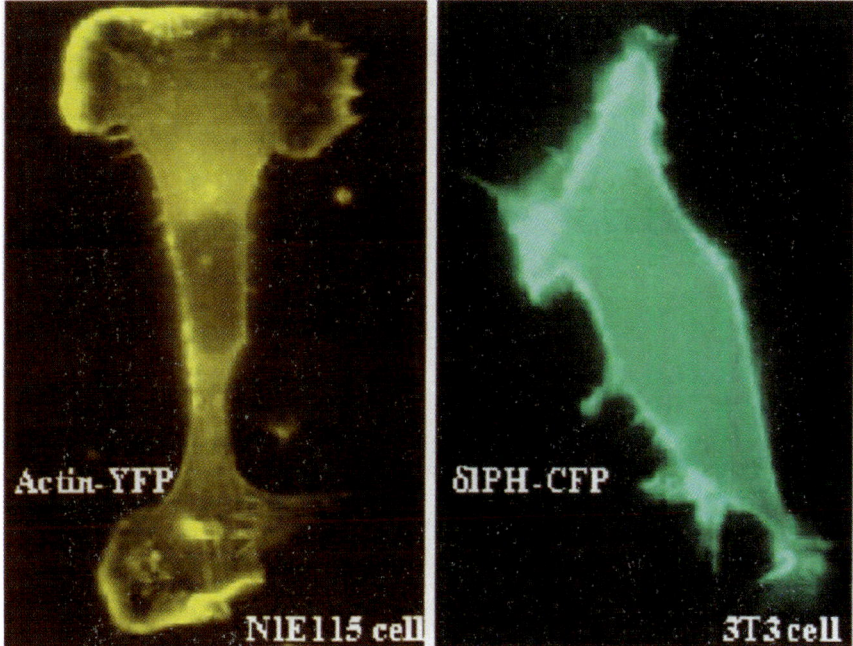

Figure 6. Localization of actin and PLC using YFP-Actin and PLCPH-CFP constructs

proteins with the most distinct emission spectra are used. One advantage of this pair is that an overlap between proteins appears yellow, so one can clearly distinguish between the individual colors and the overlap. These probes can monitor protein translocation or a change in fluorescence intensity in response to the effects of a gene or a stimulus. By making stable cell lines that express these probes, one can rapidly screen for drug candidates affecting these pathways.

GFP technology has now been forged into all fields of biology. Since then, cancer researchers have added GFP to tumors in mice to make it easier to track runaway cells and test the effectiveness of anti-metastasis drugs. Others are creating plants that glow in response to environmental pollutants, potentially saving workers from exposure to deadly toxins. Scientific research has been often likened to shining a beam of light into the darkness. Instead of just being a white light, GFP technology made this realm into a spectrum of bewildering colors.

Acknowledgments

We thank Laura Cipp, Ekta Gupta, Sanjay Patel, John Worley, and Donna Miller for their technical help over the years. National Institute of Health Grants GM 43422 and GM 60376 supported the work in the author's laboratory.

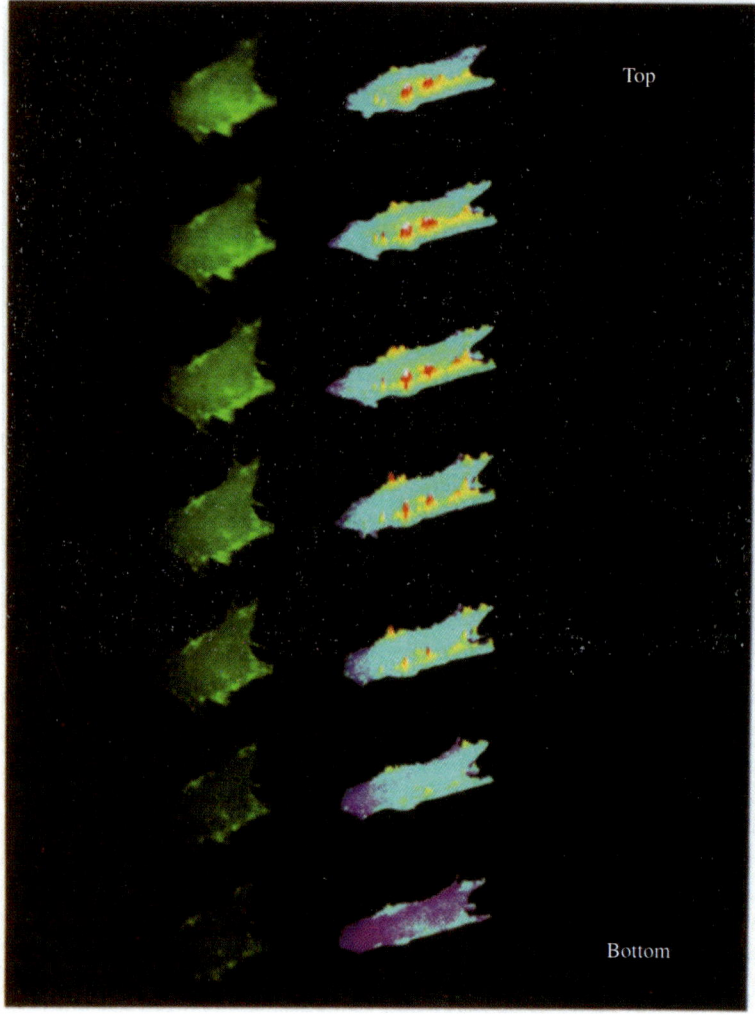

Top

Bottom

Figure 7.   Gradient distribution of PH-GFP/PIP2 in living cells

References

1. Aizawa, H., Y. Fukui and I. Yahara (1997) Live Dynamics of Dictyostelium Cofilin Suggests a Role in Remodeling Actin Latticework Into Bundles. *Journal of Cell Science* 110: 2333–2344.
2. Colbert, H.A., T.L. Smith and C.I. Bargmann (1997) Osm-9, a Novel Protein With Structural Similarity to Channels, Is Required For Olfaction, Mechanosensation, and Olfactory Adaptation in Caenorhabditis Elegans. *Journal of Neuroscience*, 17: 8259–8269.
3. Feng, X., J. Zhang, L.S. Barak, T. Meyer, M.G. Caron and Y.A. Hannun (1998) Visualization of dynamic trafficking of a protein kinase C betaII/green fluorescent protein conjugate reveals differences in G protein-coupled receptor activation and desensitization. *Journal of Biological Chemistry*, 273: 10755–62.

4. Oancea, E. and T. Meyer (1998) Protein kinase C as a molecular machine for decoding calcium and diacylglycerol signals. *Cell*, **95**: 307–18.

5. Servant, G., O.D. Weiner, E.R. Neptune, J.W. Sedat and H.R. Bourne (1999) Dynamics of a chemoattractant receptor in living neutrophils during chemotaxis. *Molecular Biology of the Cell*, 10: 1163–78.

6. Shen, K., M.N. Teruel, K. Subramanian and T. Meyer (1998) CaMKIIbeta functions as an F-actin targeting module that localizes CaMKIIalpha/beta heterooligomers to dendritic spines. *Neuron*, 21: 593–606.

7. Stauffer, T.P., S. Ahn and T. Meyer (1998) Receptor-induced transient reduction in plasma membrane PtdIns(4,5)P2 concentration monitored in living cells. *Current Biology*, 8: 343–6.

8. Tall, E.G., I. Spector, S.N. Pentyala, I. Bitter and M.J. Rebecchi (2000a) Dynamics of phosphatidylinositol 4,5-bisphosphate in actin-rich structures. *Current Biology*, 10: 743–6.

9. Venkateswarlu, K., P.B. Oatey, J.M. Tavare and P.J. Cullen (1998) Insulin-Dependent Translocation of Arno to the Plasma Membrane of Adipocytes Requires Phosphatidylinositol 3-Kinase. *Current Biology*, 8: 463–466.

10. Wang, D.S., T.L. Deng and G. Shaw (1997) Membrane Binding and Enzymatic Activation of a Dbl Homology Domain Require the Neighboring Pleckstrin Homology Domain. *Biochemical & Biophysical Research Communications*, 234: 183–189.

11. Wang, D.S., R. Miller, R. Shaw and G. Shaw (1996) The pleckstrin homology domain of human beta I sigma II spectrin is targeted to the plasma membrane in vivo. *Biochemical & Biophysical Research Communications*, 225: 420–6.

12. Essen, L.O., O. Perisic, R. Cheung, M. Katan and R.L. Williams (1996) Crystal structure of a mammalian phosphoinositide-specific phospholipase C delta. *Nature*, 380: 595–602.

13. Rebecchi, M.J. and S.N. Pentyala (2000) Structure, Function, and Control of Phosphoinositide-Specific Phospholiopase C. *Physiological reviews*, 80: 1291–1335.

14. Feng, J.F., S.G. Rhee and M.J. Im (1996) Evidence that phospholipase delta1 is the effector in the Gh (transglutaminase II)-mediated signaling. *Journal of Biological Chemistry*, 271: 16451–4.

15. Glaser, M., S. Wanaski, C.A. Buser, V. Boguslavsky, W. Rashidzada, A. Morris, M. Rebecchi, S.F. Scarlata, L.W. Runnels, G.D. Prestwich, J. Chen, A. Aderem, J. Ahn and S. McLaughlin (1996) Myristoylated alanine-rich C kinase substrate (MARCKS) produces reversible inhibition of phospholipase C by sequestering phosphatidylinositol 4,5-bisphosphate in lateral domains. *Journal of Biological Chemistry*, 271: 26187–93.

16. Grobler, J.A. and J.H. Hurley (1998) Catalysis by phospholipase C delta1 requires that Ca2+ bind to the catalytic domain, but not the C2 domain. *Biochemistry*, 37: 5020–8.

17. Kim, Y., T. Park, Y.H. Lee, K.J. Baek, P. Suh, S.H. Ryu and K. Kim (1999) PLC-delta1 is activated by capacitative calcium entry that follows PLC-beta activation upon bradykinin stimulation. *Journal of Biological Chemsitry*, 274: 26127–26134.

18. Matsushima, H., S. Shimohama, J. Kawamata, S. Fujimoto, T. Takenawa and J. Kimura (1998) Reduction of platelet phospholipase C-delta1 activity in Alzheimer's disease associated with a specific apolipoprotein E genotype (epsilon3/epsilon3). *International Journal of Molecular Medicine*, 1: 91–3.

19. Park, E.S., J.H. Won, K.J. Han, P.G. Suh, S.H. Ryu, H.S. Lee, H.Y. Yun, N.S. Kwon and K.J. Baek (1998) Phospholipase C-delta1 and oxytocin receptor

signalling: evidence of its role as an effector. *Biochemical Journal*, 331: 283–9.

20. Prasanna Murthy, S.N., J.W. Lomasney, E.C. Mak and L. Lorand (1999) Interactions of Gh/transglutaminase with PLC-delta1 and with GTP. *Proceedings of National Academy of Sciences (USA)*, 96: 11815–11819.

21. Tanino, H., S. Shimohama, Y. Sasaki, Y. Sumida and S. Fujimoto (2000) Increase in phospholipase C-delta1 protein levels in aluminum-treated rat brains. *Biochemical & Biophysical Research Communications*, 271: 620–5.

22. Rebecchi, M., V. Boguslavsky, L. Boguslavsky and S. McLaughlin (1992a) Phosphoinositide-specific phospholipase C-δ_1: effect of monolayer surface pressure and electrostatic surface potentials on activity. *Biochemistry*, 31: 12748–53.

23. Rebecchi, M.,A. Peterson and S. McLaughlin (1992b) Phosphoinositide-specific phospholipase C-delta 1 binds with high affinity to phospholipid vesicles containing phosphatidylinositol 4,5-bisphosphate. *Biochemistry*, 31: 12742–7.

24. Rebecchi, M.J., R. Eberhardt, T. Delaney, S. Ali and R. Bittman (1993) Hydrolysis of short acyl chain inositol lipids by phospholipase C-δ_1. *Journal of Biological Chemistry*, 268: 1735–41.

25. Rebecchi, M.J. and S. Scarlata (1998) Pleckstrin Homology Domain: a common fold with diverse functions. *Annual Review of Biophysics and Biomolecular Structure*, 27.

26. Scarlata, S., R. Gupta, P. Garcia, H. Keach, S. Shah, C.R. Kasireddy, R. Bittman and M.J. Rebecchi (1996) Inhibition of phospholipase C-delta 1 catalytic activity by sphingomyelin. *Biochemistry*, 35: 14882–8.

27. Tall, E., G. Dorman, P. Garcia, L. Runnels, S. Shah, J. Chen, A. Profit, Q.M. Gu, A. Chaudhary, G.D. Prestwich and M.J. Rebecchi (1997) Phosphoinositide Binding Specificity Among Phospholipase C Isozymes As Determined By Photo-Cross-Linking to Novel Substrate and Product Analogs. *Biochemistry*, 36: 7239–7248.

28 Wang, T., L. Dowal, R. El-Maghrabi, M.J. Rebecchi and S. Scarlata (2000) Plekstrin homology domain of PLC-beta2 confirs G-big activation to the catalytic core. *Journal of Biological Chemistry*, 17: 7466–9.

29. Wang, T., S. Pentyala, J. Elliot, L. Dowal, E. Gupta, M.J. Rebecchi and S. Scarlata (1999a) Selective interaction of the C2 domains of phosphoplipase C-beta1 and beta2 with activated Gq. *Proceedings of the National Academy of Sciences, USA*, 6: 7843–6.

30. Wang, T., S. Pentyala, M.J. Rebecchi and S. Scarlata (1999b) Differential Association of the pleckstrin homology domains of phospholipases C beta1, C–β_2, and C–δ_1 with lipid bilayers and the betagamma subunits of heterotrimeric G proteins. *Biochemistry*, 38: 1517–24.

31. Cifuentes, M.E., L. Honkanen and M.J. Rebecchi (1993) Proteolytic fragments of phosphoinositide-specific phospholipase C-delta 1. Catalytic and membrane binding properties. *Journal of Biological Chemistry*, 268: 11586–93.

32. Pentyala, S.N., T.C. Whyard, W.C. Waltzer, A.G. Meek and Y. Hod (1998) Androgen induction of urokinase gene expression in LNCaP cells is dependent on their interaction with the extracellular matrix. *Cancer Letters*, 130: 121–6.

33. Keating, T.J. and R.J. Cork (1994) Improved spatial resolution in ratio images using computational confocal techniques. *Methods in Cell Biology*, 40: 221.

34. Tall, E.G., I. Spector, S.N. Pentyala, I. Bitter and M.J. Rebecchi (2000b) Dynamics of phosphatidylinositol 4,5-bisphosphate in actin-rich structures. *Curr. Biol*, 10: 743–6.

35. Huang, C.F., J.R. Hepler, L.T. Chen, A.G. Gilman, R.G.W. Anderson and S.M. Mumby (1997) Organization of G Proteins and Adenylyl Cyclase At the Plasma Membrane. *Molecular Biology of the Cell*, 8: 2365–2378.

36. Lemmon, M.A., K.M. Ferguson, O.B.R, P.B. Sigler and J. Schlessinger (1995) Specific and high-affinity binding of inositol phosphates to an isolated pleckstrin homology domain. *Proceedings of the National Academy of Sciences of the United States of America*, 92: 10472–6.

37. Musacchio, A., T. Gibson, P. Rice, J. Thompson and M. Saraste (1993) The PH domain: a common piece in the structural patchwork of signalling proteins. *Trends in Biochemical Sciences*, 18: 343–8.

Radiobiology and Bio-Medical Research
Edited by K.P. Mishra
Copyright © 2004 Narosa Publishing House, New Delhi, India

20. Peptide-Membrane Interaction in a Novel Magnetically Oriented Lipid Bilayers Induced by Biologically Active Peptides as Studied by ^{31}P and ^{13}C NMR Spectroscopy

Akira Naito

Faculty of Engineering, Yokohoma National University
75-5 Tokiwadai, Hodogaya-ku, Yokohoma 240-8501, Japan

Magnetically oriented lipid bilayer systems have been recognized as a powerful means to provide detailed information on conformation and dynamics of oriented molecules in biomembranes. Although diamagnetic molecules are not normally oriented spontaneously to the magnetic field because of a very small diamagnetic anisotropy, assembly of lipid bilayers in the liquid crystalline state can give enough diamagnetic anisotropy to align to the static magnetic field. We demonstrate here that a melittin-dimyristoylphosphatidylcholine (DMPC) model membrane exhibits a new class of magnetic ordering of lipid bilayers above the gel and liquid crystalline phase transition temperature (Tm = 24 °C) as a result of induced fragmentation and fusion of membrane. We also show that this magnetically oriented lipid bilayers containing melittin can be used to investigate the secondary structure, orientation and dynamics of the peptides bound to membrane. Melittin has powerful hemolytic activity in addition to voltage-dependent ion conductance across planar lipid bilayers at low concentration. It also causes selective micellization of bilayers as well as membrane fusion at high concentration. A number of studies have been performed to determine the nature of the interaction with membrane, although there is still no consensus on the nature of its interaction with membrane, partly because many of the biophysical techniques applied to the study of protein structure and interaction are difficult to apply membrane system. We attempted here to use this magnetically oriented lipid bilayers containing melittin to investigate the structure, orientation and dynamics of melittin bound to the magnetically oriented lipid bilayers to understand the interaction of melittin with membrane [1].

Isotopically labeled melittin molecules were synthesized by an ABI peptide synthesizer using a solid phase method and purified by means of an HPLC. Peptide:DMPC (molar ratio of 1:10) was dissolved in methanol and the solvent was completely removed *in vacuo*, followed by hydration with deionized water and tris buffer (pH 7.5). ^{31}P and ^{13}C NMR spectra were recorded on a Chemagnetics

CMX 400 NMR spectrometer with and without MAS conditions under high power proton decoupling.

Figure 1 shows the 31P NMR spectra of lipid bilayer containing melittin at various temperature. When the temperature was lowered, powder pattern was

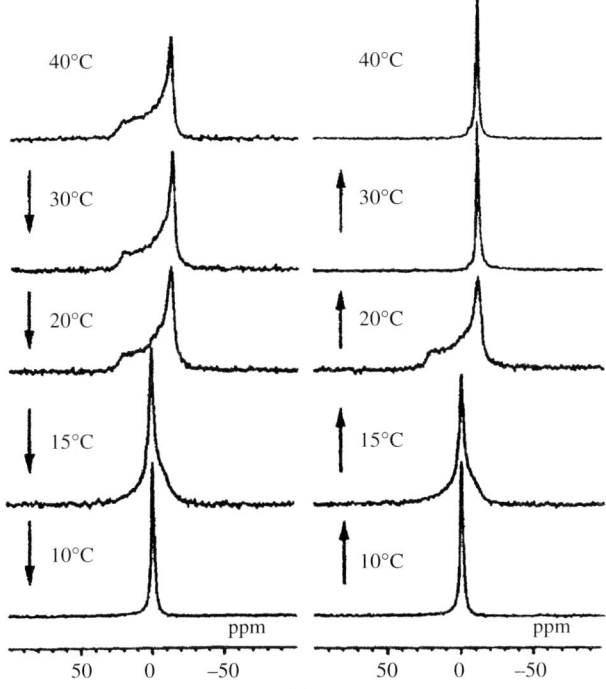

Figure 1. Temperature variation of ^{31}P NMR spectra of melittin-DMPC bilayers hydrated with deionized water. The arrows indicate the process of temperature change.

observed above 20 °C, whereas a narrow signal appeared at 10 °C. This result indicates that the fragmentation of bilayer causes formation of small sized bilayer particles exhibiting isotropic motion. When the temperature was raised, the single ^{31}P NMR line was displaced upfield by 12 ppm at 40 °C referred to that at 10 °C. This result indicates that bilayer surface is oriented parallel to the static magnetic field after the process of fragmentation and fusion. Giant vesicles with a diameter larger than 20 μm were observed using a microscope equipped with differential interference optics after the melittin-DMPC dispersion was kept at 25 °C for 3 hours. Therefore, we concluded that elongated cylindrical vesicles are formed with the long axis parallel to the magnetic field as shown in Figure 2. It is emphasized that the melittin molecules also align to the magnetic field when they strongly interact with the magnetically oriented membrane.

Figure 3 shows the ^{13}C NMR spectra of [1–^{13}C]Ile^{20}-melittin incorporated into the DMPC bilayers at –60°C and 40°C with and without MAS conditions. A broad asymmetrical powder pattern characterized by $\delta_{11} = 241$, $\delta_{22} = 189$, and $\delta_{33} = 96$ ppm appeared at –60°C. A narrowed ^{13}C NMR signal was observed at

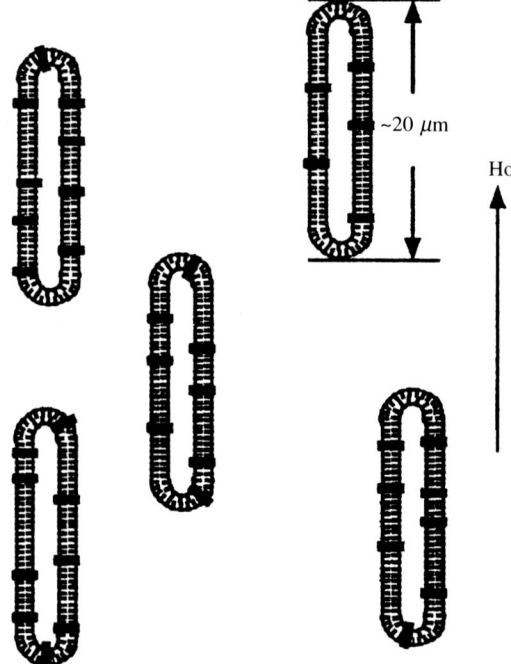

~20 μm

Ho

Figure 2. Schematic representation of the elongated vesicles of melittin-DMPC bilayers in the presence of a strong magnetic field. The longer axis is parallel to the magnetic field, and most of the bilayer surface in the vesicles is parallel to the magnetic field.

174.8 ppm for Ile[20] C = O by fast MAS experiment at 40°C, and its position was displaced upfield by 4.6 ppm in the oriented state. Since an axially symmetrical powder pattern with an anisotropy at 40°C is not broad as that at –60°C, it is evident that the α-helical segment undergoes rapid reorientation about the helical axis. The direction of the C = O is nearly parallel to the axis of α-helix because Ile[20] is involved in α-helix. The observed chemical shift of [1–^{13}C]Ile[20]-melittin in the magnetically oriented bilayers was close to the value of $\delta_\perp$ component of the ^{13}C chemical shift tensor of the Ile C=O in α–helix structure. It is, therefore, evident that the axis of α-helix is perpendicular to the magnetic field, namely parallel to the bilayer normal. We have further measured the ^{13}C NMR spectra of [1-^{13}C] Gly^3, [1-^{13}C]Val^5, [1-^{13}C]Gly^{12}, and [1-^{13}C]Leu^{16}-melittin. ^{13}C resonances are located at the $\delta_\perp$ positions on the respective powder patterns. However, the powder patterns were quite different from each other, indicating that α-helix is not rotating about the helical axis, but rotating about the axis tilting to the helical axis and this symmetric axis should be parallel to the bilayer normal. Detail analysis of the powder patterns indicate that the N– and C– terminal helical rods are tilting 30° and 10° to the symmetric axis, respectively.

It is of interest to discuss the dynamic structure of melittin in a lipid bilayer in relation to its lytic activity. The present results indicate that melittin forms the

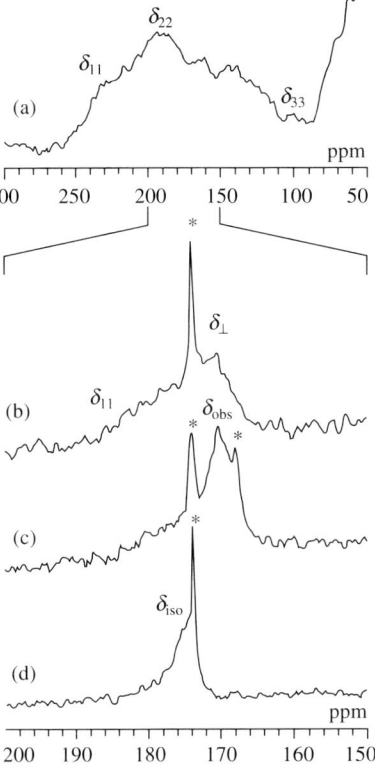

Figure 3. **Temperature variation of ^{13}C NMR spectra of a DMPC bilayer in the presence of [1-^{13}C]Ile^{20}-melittin in the static condition at –60 °C (a) and slow MAS (b), static (c), and fast MAS (d) conditions at 40 °C. The signals marked by asterisks are assigned to the C=O groups of DMPC.**

transmembrane α-helix in the lipid bilayer, whose symmetric axis is parallel to the bilayer normal. It was also shown that the transmembrane α-helix is not static, but the N- and C-terminal α-helical rods rotate or reorient rapidly about the symmetric helical axis. Although the direction of the symmetric α-helical axis is parallel to the bilayer normal, the local helical axis may precess about the bilayer normal by making angle of 30° and 10° for the N- and C-terminal helical rods, respectively as shown in Figure 4a. This dynamic behavior of melittin strongly suggests that it exists as a monomer in the lipid bilayer at temperatures higher than the Tm. This monomeric transmembrane helix of melittin is considered to be unstable in the lipid bilayers because of the amphiphilic nature. These unstable helices might associate to make pores by aligning the hydrophobic side towards the lipid bilayers. The lytic activity of melittin towards the lipid bilayer can be explained in terms of spontaneous association of melittin in the lipid bilayer, resulting in the formation of pores in the lipid bilayer surface to separate the lipid bilayers and, consequently, in the complete fragmentation into disc-type micelles as shown schematically in Figure 4b.

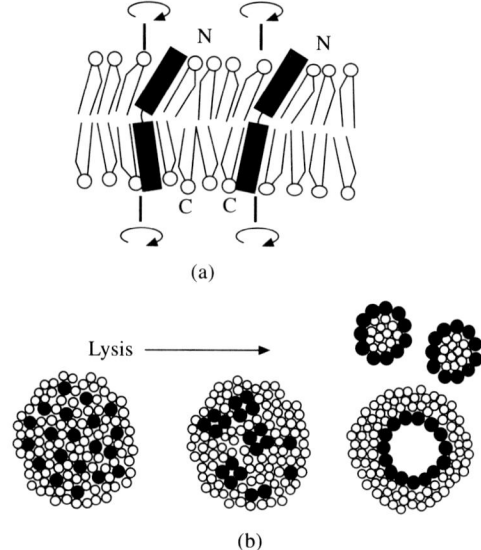

(a)

(b)

Figure 4. (a) Schematic representation of the orientation of melittin helices bound to magnetically oriented lipid bilayers. N- and C-terminal helix axes make angles of 30° and 10°, respectively, with the symmetric axis, which is perpendicular to the bilayer surface. (b) The lytic process of lipid bilayers in the presence of melittin at temperature below Tm. o, $\sum$, lipid and melittin molecules, respectively.

References

1 Naito, A., T. Nagao, K. Norisada, T. Mizuno, S. Tuzi and H. Saito, (2000) *Biophysical Journal* 78: 2405–2417.

Radiobiology and Bio-Medical Research
Edited by K.P. Mishra

21. Initiation of Forward Motility in the Goat Caput Epididymal Immature Sperm *in vitro*

G.C. Majumder and B.S. Jaiswal

Indian Institute of Chemical Biology, 4 Raja S.C. Mullick Road, Jadavpur,
Calcutta–700 032, India

Introduction

Mammalian spermatozoa after formation in testis acquire forward motility (FM) potential during transit through epididymis. Biochemical basis of this epididymal sperm maturation is largely unknown. To gain an insight into the mechanism of this motility initiation, it is essential to develop an *in vitro* model system that will permit induction of FM in the immature caput-epididymal sperm. Hoskins and his associates (1–2) were first successful to induce FM in bovine sperm derived from caput epididymis upon incubation with a media containing seminal plasma and theophylline, an inhibitor of cyclic phosphodiesterase. In this *in vitro* model, both theophylline and seminal plasma are essential for motility initiation. Theophylline is believed to increase intrasperm cyclic AMP level by inhibiting cAMP phosphodiesterase. Bicarbonate is believed to activate motility by stimulating sperm adenylate cyclase activity thereby enhancing the intrasperm level of cAMP.

Goat Epididymal Sperm Model for Motility Initiation

We have developed a goat caput-sperm model to investigate biochemical basis of sperm motility initiation [3]. Investigation was carried out to elucidate the functional inter-relationship among exogenous FMP, bicarbonate and pH for the initiation of sperm motility *in vitro*. As shown in Table 1, FM is induced approximately 55 percent of spermatozoa when incubated in modified Ringers solution (pH 8.0) containing theophylline, dialysed EP and bicarbonate. Theophylline is essential for FM initiation by EP and bicarbonate, but alone it fails to initiate FM in sperm cells.

Purification of Forward Motility Protein

Hoskins *et al.* [2, 1] have shown that the active principle of bovine seminal plasma is a heat-stable protein designated as forward motility promoting protein and it has been partially purified. We have, for the first time, purified motility initiating protein (MIP) to apparent homogeneity from epididymal plasma (goat)

Table 1. Effect of different reagents on initiation of motility in the goat caput-sperm.

Caput-sperm were incubated with the indicated additions for 10 min in RPS medium, pH 8.0 containing antisticking factor (ASF, 250 μg protein mL^{-1}) and induced forward motility (FM) was measured as described earlier (3). The data shown are mean $\pm$ s.e.m. of six experiments. $P < 0.05$ in case of a v.c and b v.c and insignificant ($P > 0.05$) in case of a v.b.

Additions	Forward motility (%)
Control	0
+ 30 mM theophylline	0
+ 30 mM theophylline + EP (0.6 mg protein ml^{-1})	38 $\pm$ 5[a]
+ 30 mM theophylline + 25 mM bicarbonate	44 $\pm$ 3[b]
+ 30 mM theophylline + EP (0.6 mg protein ml^{-1})	
+ 25 mM bicarbonate	54 $\pm$ 3[c]
+ 25 mM bicarbonate + EP (0.6 mg protein ml^{-1})	0

by using multiple biochemical fractionation procedures, namely, ammonium sulphate fractionation, DEAE-cellulose chromatography, concanavalin A-Sepharose affinity chromatography, chromatofocussing and Sephacryl S-200 gel filtration (5–7).

The purity of the isolated MIP has firmly been established by using several modern analytical techniques such as polyacrylamide gel electrophoresis, HPLC and isoelectricfocussing.

Properties of MIP

Major characteristics of the purified MIP have been summarized in Table 2. The molecular mass of the native MIP as determined by Sephacryl S-200 gel filtration, HPLC and native PAGE is approx 125 KDa. It is a dimeric protein of about 70 and 54 KDa subunits. The factor has high protein specificity and affinity for inducing FM in the immature epididymal sperm *in vitro*. MIP at 30 μg/ml level showed maximal motility-promoting activity. MIP is also capable of enhancing FM of the goat mature cauda-sperm. Ca^{2+} and Mg^{2+} stimulate MIP activity. Treatment of spermatozoa with the motility promoter caused a significant increase of the intrasperm level of cyclic AMP thereby suggesting that FMP stimulates sperm motility by activating membrane-bound adenyl cyclase. It is an acidic protein with isoelectric point (PI) of about 4.75. It is stable to heat treatment at 100°C for three minutes. It is a glycoprotein that binds with high affinity to concanavalin A. It contains mannose, galactose and N-acetyl glucosamine approximately in the ratios of 6:1:6. MIP markedly loses its activity when incubated with α-mannosidase, β-N-acetyl glucosaminidase and proteolytic enzymes indicating that both the sugar and protein parts are essential for its biological activity. Immunofluorescence studies show that MIP is localized on the outer surface of the sperm with special reference to the head region. MIP is strongly immunogenic. Antibody against it markedly inhibits FM of mature goat sperm as well as the MIP-induced motility initiation *in vitro* in the immature sperm thereby implicating that MIP has potential for use as a contraceptive vaccine for control of population growth. Using enzyme-linked immunosorbent assay (ELISA) the distribution of MIP has been analysed in a variety of tissues and in some body fluids. The specific activity of the motility initiator is highest in epididymal plasma. FMP or/and immunologically cross-reactive protein(s) are present in significant level in bone marrow and blood serum. The other tissues tested have low/insignificant levels of MIP. The factor occurs in sperm plasma membrane and the membrane-bound MIP level increases markedly during the epididymal sperm maturation. The data show that the purified protein (MIP) from goat EP is a physiological motility-activating protein.

Table 2. Major characteristics of motility initiating protein (MIP)

- MIP is a 125 KDa protein with 2 subunits—70 KDa and 54 KDa
- A glycoprotein
- PI—4.80
- High protein specificity
- Maximal activity at approx. 30 μg/ml (250 mM)
- Sugar part essential for activity
- Activated (~50%) by 5 μM Ca^{++}
- Strongly antigenic
- MIP-receptors are localized on sperm head
- Antibody treatment causes loss of sperm motility
- EP—richest source of MIP
- MIP also present in sperm membrane
- MIP acts by elevating sperm cAMP level
- MIP also stimulates mature cauda-sperm motility

References

1. Acott, T.S. and D.D. Hoskins, (1978) *J. Biol. Chem.* 253: 6744–6750.
2. Hoskins, D.D., H. Brandt and T.S. Acott, (1978) *Fed. Proc* 37: 2534–2542.
3. Jaiswal, B.S. and G.C. Majumder, (1998) *Reprod. Fertil. Dev.* 10: 299–307.
4. Majumder, G.C. and B.S. Jaiswal, (1998) Indian Patent Application No. 505/DEL/98.
5. Majumder, G.C. and B.S. Jaiswal, (1998) USA Patent Application No. US09/203093.
6. Majumder, G.C. and B.S. Jaiswal, (1998) European Countries Patent Application No. 983099086–2105.
7. Vijayraghavan, S., L.M. Critchlow and D.D. Hosmins, (1985) *Biol. Reprod.* 32: 489–500.

Radiobiology and Bio-Medical Research
Edited by K.P. Mishra
Copyright © 2004 Narosa Publishing House, New Delhi, India

22. Probing the Molecular Mechanisms of Enzyme Catalysis and Regulation

Gotam K. Jarori

Department of Biological Sciences, Tata Institute of Fundamental Research,
Colaba, Mumbai -400005, India

Abstract: Enzymes are highly specific, efficient biocatalysts whose catalytic potential can be modulated by interaction with regulatory ligands. For elucidation of molecular basis of enzyme catalysis, knowledge of enzyme structure and its complexes with ligands is required. In general, for macromolecules (MW > 30 kDa) x-ray crystallography is the only method capable of providing detailed information about the 3D-structure. However, crystallographic information on the conformation of the substrates at the active site is sparse in comparison, because of difficulties associated with co-crystallizing enzyme-substrate complexes. NMR methods offer advantage in that the measurements can be made in solution and on several different complexes to derive the conformation of enzyme-bound ligands.

Introduction

In this talk, I will describe our studies on determination of conformation of enzyme bound nucleotides in some of the metabolic kinases. Ligand induced conformational changes, relevant to catalysis and regulation of enzyme activity as probed by fluorescence spectroscopy will also be discussed.

Two striking features of enzyme catalyzed reactions are their exquisite specificity and potential to enormously enhance the rate of a reaction. Early on, it was realized that enzymes first form a specific complex (Michaelian complex) with substrates and this specific arrangement of substrates at the active site ensures the specificity of the reaction catalyzed. The observed enhancement of rate by enzymes implies that activation energy barrier along the reaction path are lowered considerably as compared to the uncatalyzed reaction. In 1946, Linus Pauling proposed that the lowering of the activation energy in enzyme catalysis stems from enzyme's affinity for the reaction transition-state (TS), exceeding its affinity for the substrates [17, 18]. This proposal has great popular appeal and led to the development of transition state inhibitors and antibody-enzymes (abzymes) [8, 22]. Recent investigations, however, have brought into focus the contribution of thermal motions and ground state conformers associated with the Michaelis complex to the rate of the enzymatic reactions [2]. Thus a precise knowledge of conformation of substrates bound at the active site of the enzyme in Michaelian complex is critical for our understanding of the molecular mechanism of catalysis. In this article, we describe some of our studies on a group of metabolic kinases where the conformation of enzyme-bound nucleotide (ATP and ADP) have been determined using solution NMR studies. Every kinase catalyzes the transfer of γ-phosphoryl group of ATP to an acceptor substrate. The reaction requires a Mg(II) as an obligatory cofactor, which can be substituted by paramagnetic cations such as Mn(II) and Co(II).

Free ATP molecule has three distinct internal mobilities, namely, those associated with the flexible phosphoryl chain, the rotation of the adenine base about glycosidic bond, and the sugar pucker. Assuming that the cation-nucleotide complex has a unique conformation when bound to the enzyme, a determination of this conformation will also reveal how the internal motions are arrested in the bound species. We are using high resolution NMR spectroscopy to obtain the bound ligand conformation. Our experimental strategy involves determination of location of cation with respect to phosphate chain on the basis of distance dependent paramagnetic contribution to the spin-lattice relaxation times (T1) of ^{31}P nuclei [4, 5] and inter-proton distances on the basis of Transferred Nuclear Overhauser Effect Spectroscopy (TRNOESY) [14, 15, 6, 7, 9, 10, 11].

Experimentally, paramagnetic effects on nuclear spin-relaxation are measured on samples in which the cation concentration is a small fraction of the ligand and observed relaxation is altered by exchange between the diamagnetic and paramagnetic complex. The paramagnetic contribution, T_{1p}^{-1} is given by

$$T_{1p}^{-1} = T_{\text{lobsd}}^{-1} - T_{1D}^{-1} = \{p/(T_{1M-\tau_M})\}$$

with p = [Cation]/[Ligand] and

$$T_{1M}^{-1} = \{(C/r)^6 f(\tau_c)\}$$

where T_{1D}^{-1} is the relaxation rate in diamagnetic complex and τ_M is the life time of the paramagnetic complex, C is constant for a given cation (Mn(II) or Co(II)) and the relaxing nucleus (^{1}H /^{31}P/^{13}C/^{15}N), τ_c is the correlation time for the dipolar interaction and r is the distance between cation and the relaxing nucleus [13, 21]. A typical experimental T_1 measurement on ^{31}P of ATP in creatinekinase.ATPCo(II) complex is shown in Figure 1A.

In a TRNOESY experiment, the inter-proton distances among various protons of a ligand in a macromolecular complex are obtained by making NOE measurements when the ligand is in fast exchange between the bound and the free states. A typical TRNOESY spectrum (yeast hexokinase. ADPMg(II) complex; [enzyme]:[ADP] ratio 1:10) for enzyme-nucleotide complex is shown in Figure 1 (B). Since the rotational correlation time of the bound complex is much longer (by a factor of 10^3–10^4) than that of the free ligand, the cross relaxation rates in the bound state are correspondingly larger. Under fast exchange conditions, inter-proton distances can be determined by complete relaxation matrix analysis of time dependent TRNOE build up data [3, 16].

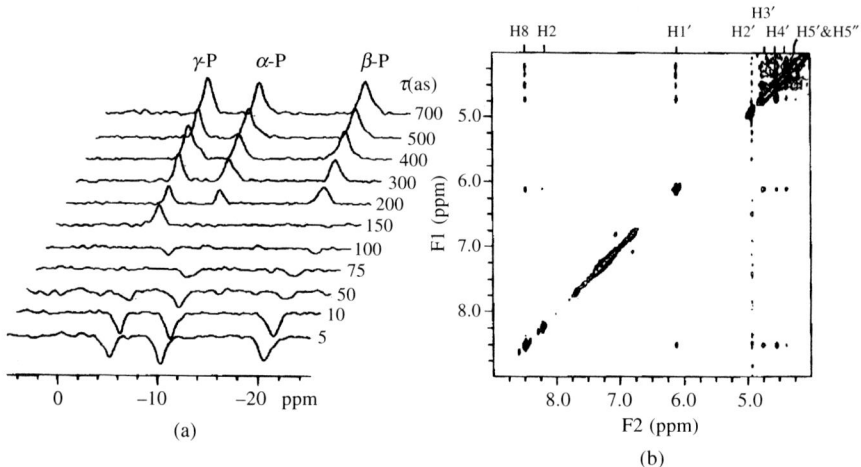

Figure 1. (A) A typical T1 measurement. (B) A typical 2D TRNOE spectrum.

Coordination State of Cation in Enzyme-Metal-Nucleotide Complexes

Extensive (temperature and frequency dependent) measurements of ^{31}P-relaxation rates in enzyme-bound nucleotides in the presence of Mn(II) showed that the relaxation rates are exchange limited, that is, $T_{1M}^{-1} \gg \tau_M^{-1}$ and hence T_{1D}^{-1} is bereft of any structural information. Using Co(II), which induces weaker nuclear relaxation than Mn(II), it could be shown for nucleotide.metal complexes of creatine kinase (CK) and arginine kinase (AK) that the metal ion forms a first coordination complex (see Table 1).

Table 1. Cation–^{31}P distances in various enzyme complexes.

Complex	Co(II)-^{31}P Distance (r) (Å)			Reference
	α–P	β–P	γ–P	
CK.ADPCo(II)	3.0 – 4.0	2.9 – 3.8		Jarori et al., 1995.
CK.ATPCo(II)	3.2 – 4.2	3.0 – 4.0	2.9 – 3.8	Jarori et al., 1995.
AK.ADPCo(II)	3.2 – 4.2	3.4 – 4.5		Jarori et al., 1989.
AK.ATPCo(II)	3.5 – 4.6	3.2 – 4.2	3.2 – 4.2	Jarori et al., 1989.

Glycosidic Torsion (χ) and Sugar Pucker

TRNOESY measurements were made to determine interproton distances in the adenosine moieties of nucleotides in complex with several different proteins, namely, creatine kinase [14], arginine kinase [15], pyruvate kinase [6], phosphoribosyl pyrophosphate synthetase [7], hexokinase [10, 11]. Interproton distances obtained by full relaxation matrix analysis of TRNOE data were used as restraints in molecular modeling. The adenosine conformations (glycosidic torsion) and ribose puckers corresponding to the energy minimized structures compatible with the distances obtained from NOE data are given in Table 2. The glycosidic torsion angles fall in a rather narrow range of 51 ± 8°. Observed similarity in conformation of adenosine moiety reflects similarity in the 3D structure of nucleotide binding domains in different enzymes.

Table 2. Bound nucleotide conformation in various enzyme complexes

Complex	χ (deg)	Sugar Pucker	Reference
HK.ADPMg	68	${}^{\circ}T_1$	Maity & Jarori, 1997.
HK.Glc.ADPMg	52	${}^{\circ}T_1$	Maity & Jarori, 1997.
HK.Glc.ADPMg. NO_3^-	49	${}^{\circ}T_1$	Maity & Jarori, 1997.
AK.ADPMg	51	${}^{\circ}E$	Murali et al., 1994.
AK.Arg.ADPMg. NO_3^-	52	${}_1E{-}{}_1^2T$	Murali et al., 1994.
AK.ATPMg	50	${}_1E{-}{}_1^2T$	Murali et al., 1994.
CK.ADPMg	51	${}_4^0T$	Murali et al., 1993.
CK.ATPMg	51	${}_4^0T$	Murali et al., 1993.
PK.ATPMg(active site)	44	${}_4^3T{-}{}_4E$	Jarori et al., 1994.
PK.ATPMg(ancillary site)	46	${}_1E{-}{}_1^2T$	Jarori et al., 1994.
PRPPS.ATPMg	50	${}_1^0T{-}{}_1E$	Jarori et al., 1995.
Free ATP	5	${}^3E{-}{}_4^3T$	

Orientation of Phosphoryl Chain

Once the location of the cation with respect to phosphate chain, and the conformation of the adenosine moiety are known, the orientation of phosphate chain with reference to adenosine can be measured through T_{1P} measurements on ^{13}C labeled nucleotides bound to the enzyme. Such measurements were made on creatine kinase complexes of $[2-^{13}C]ATP$ and $[2-^{13}C]ADP$, which yielded distances of 10 ± 0.5 Å and 8.6 ± 0.5Å between the cation and the labeled ^{13}C, respectively [19]. These distances alone are not adequate to precisely determine the orientation of the phosphate chain. However, they do significantly reduce the allowed ranges for the possible conformations.

Recently, we attempted to determine the orientation of the phosphoryl chain in two different complexes of yeast hexokinase, namely, HK.AMPPCPMg and HK.Glc.AMPPCPMg. AMPPCP is a non-hydrolyzable analogue of ATP where the bridge oxygen between β and γ phosphoryl group is replaced by a $>CH_2$ group. Observation of NOE's between the bridge $>CH_2$ protons and ribose moiety has allowed to define the conformation of the phosphoryl chain. Results show that glucose induced conformational changes in yeast hexokinase alter the conformation in the triphosphoryl chain region of the enzyme bound nucleotide (Fig. 2) (Maity & Jarori unpublished results). Attempts are being made to measure paramagnetic relaxation in enzyme-bound equilibrium mixtures in order to probe the structural alterations in the reaction complex accompanying the enzyme turnover [20].

Regulation of Yeast Hexokinase Activity

Hexokinase catalyzes the first irreversible step in utilization of glucose. Trehalose 6-phosphate (T6P) has been shown to be a strong inhibitor of this enzyme [1]. Yeast hexokinase exist in equilibrium between two conformational states, namely 'Open' and 'Closed'. Glucose binds to 'Closed' state, which is the active form of the enzyme. We have performed Time-Resolved and Steady State fluorescence studies on interaction of T6P with hexokinase. Our studies show that T6P binds to 'Open' conformation of the enzyme. Thus T6P inhibits the enzyme by binding to 'Open', inactive form of the enzyme rather than competing with glucose to bind to 'Closed' active form of the enzyme [12].

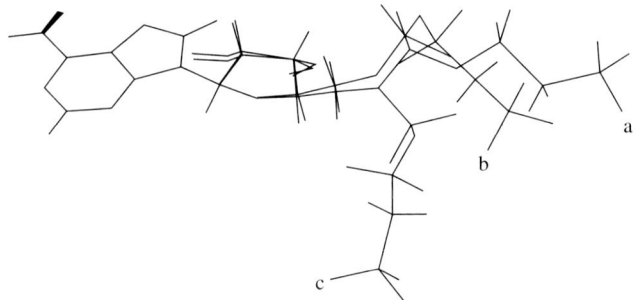

Figure 2. Schematic representation of AMPPCPMg conformation in (a) extended form, (b) HK.AMPPCPMg, and (c) HK.Glc.AMPPCPMg complex

Acknowledgements

I am grateful to Dr. H. Maity who has significantly contributed to this work. Some of the work presented here was carried out in collaboration with Dr. B. D. Ray and Dr. B. D. Nageswara Rao of Dept. of Physics, IUPUI, Indianapolis, USA.

References:

1. Blaquez, M.A., R. Lagunas, C. Gancedo and J.M. Gancedo (1993) FEBS Letts. 329: 51–54.
2. Bruice, T.C. and S.J. Benkovic (2000) *Biochemistry* 39: 6267–6274.
3. Campbell, A.P. and B.D. Sykes (1991) *J. Magn. Res.* 93: 77–92.
4. Jarori, G.K., B.D. Ray and B.D. Nageswara Rao (1985) *Biochemistry* 24: 3487–3494.
5. Jarori, G.K., B.D. Ray and B.D. Nageswara Rao (1989) *Biochemistry* 28: 9343–9350.
6. Jarori, G.K., N. Murali and B.D. Nageswara Rao (1994) *Biochemistry* 33: 6784–6791.
7. Jarori, G.K., N. Murali, R.L. Switzer and B.D. Nageswara Rao (1995) *Eur. J. Biochem.* 230: 517–524.
8. Kirby, A.J. (1996) *Acta Chem. Scand.* 50: 203–210.

9. Maity, H. and G.K. Jarori (1996) *Curr. Sci.* 71: 906–914.
10. Maity, H. and G.K. Jarori (1997) *Eur. J. Biochem.* 250: 539–548.
11. Maity, H. and G.K. Jarori (1998) *Physiol. Chem. Phys. & Med. NMR* 30-49–62
12. Maity, H., N. Maity and G.K. Jarori (2000) *J. Photochem. Photobiol. B: Biol.* 55: 20–26.
13. Mildvan, A.S. and R.K. Gupta (1978) *Methods Enzymol* 49G: 322–359.
14. Murali, N., G.K. Jarori, S.B. Landy and B.D. Nageswara Rao (1993) *Biochemistry* 32: 12941–12948.
15. Murali, N., G.K. Jarori and B.D. Nageswara Rao (1994) *Biochemistry* 33: 14227–14236.
16. Ni, F. (1994) *Prog. NMR Spectr.* 26: 517–606.
17. Pauling, L. (1946) *Chem. Eng. News* 24: 1375.
18. Pauling, L. (1948) *Nature* 161: 707–709.
19. Ray, B.D., M.H. Chau, W.K. Fife, G.K. Jarori and B.D. Nageswara Rao (1996) *Biochemistry* 35, 7239–7246.
20. Ray, B.D., G.K. Jarori and B.D. Nageswara Rao (2000) NMR determination of Structural Changes in Enzyme-Bound Reaction Complexes Accompanying Enzyme Turnover, abstract , pp. 337, XIX International Conference on Magnetic Resonance in Biological Systems (Held at Florence, Italy, Aug. 20–25, 2000).
21. Villafranca, J.J. (1984) *Phosphorus-31 NMR: Principles and Applications* Editor D.G. Gorenstein, New York: Academic Press pp. 155-174.
22. Wentworth. P. and Janda, K.D. (1998) *Curr. Opin. Biol.* 2: 138–144.

Radiobiology and Bio-Medical Research
Edited by K.P. Mishra
Copyright © 2004 Narosa Publishing House, New Delhi, India

23. Trends in Applications of EMR Spectroscopy in Biology, Life and Medical Sciences

Czeslaw Rudowicz

Department of Physics and Materials Science, City University of HK,
Kowloon Tong, Hong Kong SAR

Abstract.   The aims of this paper are (i) to provide an overview of selected topics, (ii) to focus of the latest developments and (iii) to identify the trends in applications of electron magnetic resonance (EMR) in biology, life and medical science. For the benefit of wider audience, first we provide in the nutshell some fundamentals of EMR phenomenon and an overview of EMR spectroscopy. Recent applications of EMR in the areas in question are reviewed based on an extensive survey of literature. Four major topics have been selected for a detailed presentation, namely, (1) EMR of transition ions in bio-systems, (2) high-magnetic field and high-frequency EMR of bio-systems, (3) EMR of nitric oxide and its biology, and (4) EMR and photosynthesis research. The emerging trends in EMR applications and their potential for enhancing biomedical research and technologies have been outlined. Particular focus is on applications of novel EMR techniques to solve problems in biology, life and medical sciences.

Introduction

The scope of applications of electron magnetic resonance (EMR) spectroscopy in biology, life and medical sciences is beyond what can be covered within the present page limit. Hence, in this paper we focus on the emerging trends concerning the four topics selected for detailed consideration, whereas in the invited talk [51] we have attempted to briefly cover more aspects. A full review of the area in question will be published elsewhere. The purpose of this paper is to provide researchers, who are not experts and either are interested in EMR or want to apply the EMR techniques in their own work, with: (i) some fundamentals of EMR phenomenon (Section: EMR Phenomenon), (ii) an overview of EMR spectroscopy (Section: Overview of EMR Spectroscopy), and (iii) an overview of the potential and capabilities of the EMR techniques as well as the emerging trends in EMR applications (Section: Major Topics and Trends). As much as possible, we provide pointers to the most recent specialized literature for in-depth coverage.

The scope of the survey of literature dealing with applications of EMR in biology, life and medical sciences, carried out for this work includes major review articles and books published in the last decade as well as the important relevant papers published within the last few years. To provide an overview of the latest developments, especially in the Asia-Pacific region, concerning the applications of EMR in biology, life and medical sciences, the papers presented at the First (APES'97) and Second (APES'99) Asia-Pacific EPR/ESR Symposium held at the CityU in Hong Kong in January 1997 and at the Zhejiang University in Hangzhou in October/November 1999, respectively, have also been surveyed. The topics covered at APES'97 [47, 49] and APES'99 [61] will be categorized in the full review.

Looking into the future, let us note here that several specialized conferences pertinent for this review are to be held in 2001 (see, *EPR Newsletter* vol. 11, no. 3, 2000). Information is available on the following websites: *http://www.cf.ac.uk/esr/norwich/bristol.html* for the 34th Annual International Meeting of the ESR Group of the Royal Society of Chemistry, *http://www.ismrm* for the Joint Annual Meeting of International Society for Magnetic Resonance in Medicine and ESMRMB, *http://spinchem.riken.go.jp* for the VIIth International Symposium on *Magnetic Field and Spin Effects in Chemistry and Related Phenomena*, *http://www.tau.ac.il/chemistry/ismar.html* for the International Society of Magnetic Resonance (ISMAR) Meeting, *http://www.cerm.unifi.it.icbic/icbic10.html* for the 10th International Conference on Bioorganic Chemistry, *http://www.dartmouth.edu/~eprctr/workshop2001* for Workshop on EPR Studies of Viable Biological System (especially *in-vivo*) and Related Techniques (especially oxiometry), *http://iris1.chemie.uni-kl.de/spin2001.html* for the 3rd International Conference on Nitroxide Radicals, *Spin: Synthesis, Properties and Implications of Nitroxides*, *http://www.mol.uj.edu.pl/EPRWorkshop* for the 5th Workshop on *Recent Advances in Applications of EPR in Biology and Medicine*, *http://www.ied.edu.hk/has/phys/apepr* for the 3rd Asia-Pacific EPR/ESR Symposium.

EMR Phenomenon

In brief, the phenomenon of electron magnetic resonance (EMR) can be described as the observation of transitions induced by an oscillating source of electromagnetic radiation ($\mathbf{B_1}$) between the energy levels of a system of electronic magnetic dipoles, described equivalently as a '*spin*' system. The transitions are observed most commonly between the energy levels of the system split by an external static magnetic field ($\mathbf{B_0}$), the splitting being referred to as the Zeeman splitting. Over the years, the frequency range of the electromagnetic radiation inducing EMR transitions has been extended from the microwave (~9.5 GHz) at the onset of EPR to, nowadays, submillimeter-wave range W-band (~94 GHz) and beyond, see Section 4.2. In general, the '*spin*' may be solely due to (i) the (*real*) spin angular momentum of an electron ($\mathbf{s}$), or (ii) total (*paramagnetic*) spin angular momentum ($\mathbf{S} = \Sigma \mathbf{s_i}$) of a system of unpaired electrons, or (iii) total angular momentum ($\mathbf{j} = \mathbf{l} + \mathbf{s}$ or $\mathbf{J} = \Sigma \mathbf{j_i} = \mathbf{L} + \mathbf{S}$), which includes both spin ($\mathbf{s}$ for a single spin or $\mathbf{S} = \Sigma \mathbf{s_i}$ for several spins) and orbital angular momenta ($\mathbf{l}$ for a single electron or $\mathbf{L} = \Sigma \mathbf{l_i}$ for several electrons). The physical nature of a given '*spin*' system determines the origin of the observed spectrum, and thus determines the name applicable to a particular resonance technique, for example, NMR (nuclear magnetic resonance) for *'nuclear spin'* systems, whereas EMR for either individual or collective *'electronic spin'* systems.

The term EMR is now more often used, and encompasses both electron paramagnetic resonance (EPR) and electron spin resonance (ESR), and related magnetic-resonance (MR) techniques listed in Section on Overview of EMR Spectroscopy, which all deal with individual *'electronic spin'* systems, as well as, ferromagnetic resonance (FMR) and antiferromagnetic resonance (AFMR), which deal with collective *'electronic spin'* systems. In the early years of microwave spectroscopy the term EPR was dominant, signifying the paramagnetic nature of the majority of the electronic *'spin'* systems studied. Later, when the EPR technique became widely used in chemistry and other areas outside physics, the term ESR was used in some areas, being appropriate to free radicals, for example, odd-electron species and carbon-based organic free radicals, as well as conduction electrons, where *'spin'* is due solely to the electronic spin of the unpaired electron(s), the orbital angular momentum being absent. However, the name ESR is not appropriate in general for the transition ions ($3d^n$, $4d^n$, $5d^n$; $4f^n$, $5f^n$ series) in paramagnetic crystals characterized by a non-zero total orbital angular momentum in the ground state, for which the name EPR better reflects the paramagnetic nature of the species participating in resonant absorption. The name EMR is gaining a wider acceptance [50] and is a more appropriate generic name to use, for the various MR techniques, than either EPR or ESR, being semantically symmetrical to NMR.

Overview of EMR Spectroscopy

EMR spectroscopy is now a mature technique with applications in physics, chemistry, earth sciences (geology), materials science, and, more recently, in archaeology, biology, biophysics, medical and environmental sciences. The systems studied by EMR range from minerals, biological molecules, free radicals, defects in the transition-metal and rare-earth ion crystals, optoelectronic and laser materials, semiconductors to high-temperature superconductors. Concerning the method, EMR techniques can be subdivided into (i) cw (continuous wave) EMR, (ii) pulsed EMR, and (iii) Fourier-transform (FT) EPR. The following frequency bands and ranges have been developed over the years. The early conventional cw EMR spectroscopy has evolved from the most common frequency band X (~9.5 GHz) to K (~23 GHz) and Q (~35 GHz), and today it includes also the lower bands: L (~1.5 GHz), S (~3.0 GHz) and C (~6.0 GHz), as well as the higher bands: V (~50 GHz), W (~95 GHz), far-infrared, and broad bands encompassing THz frequencies (for references, see). Most recently the multi-frequency EMR techniques, which utilize several frequencies, have become more widely used due to the developments in EMR spectrometers.

A wide range of related EMR techniques has been developed: APR (acoustic paramagnetic resonance), ZFR (zero-field EPR), electron-nuclear double resonance (ENDOR), electron-electron double resonance (ELDOR), ODMR (optically-detected magnetic resonance), electric field EPR (EFEPR), electron-spin-echo modulation (ESEM) *or* electron-spin-echo envelope modulation (ESEEM), high-frequency (HF) and high-magnetic field (HMF) EPR, and thermally-detected

(TD) EPR. Several specialized EMR applications exist, namely, (i) ESR dating and dosimetry, (ii) *in-vivo* EPR, and (iii) EPR imaging and microscopy (for references, see, 51). Description and classification of the information content obtainable from the major EMR techniques with specific references will be provided in the full review. For completeness it is worth listing the related spectroscopic and magnetic techniques, which provide complementary information: magnetic susceptibility, magnetic anisotropy studies, Mössbauer spectroscopy, optical absorption spectroscopy, and infrared spectroscopy. The basic theoretical concept underlying EMR and the related techniques, that is, the **spin Hamiltonian**, has been reviewed in detail by Rudowicz (1987) and Rudowicz & Misra (2001), whereas the operator and parameter notations used in spin Hamiltonians by Rudowicz (1987) and Rudowicz et al (2001).

Major Topics and Trends

The reasons for selection of the topics discussed below are: (i) these are the major current topics, (ii) there is a significant role of the EMR techniques, (iii) the space constraints. Note that some overlap between the topics is unavoidable, whereas other topics to be dealt with in the full review.

EMR of transition metal ions in bio-systems

Mononuclear paramagnetic metal sites in biomolecules may contain the following transition metal (TM) ions: $V^{2+}(3d^3)$, $V^{3+}(3d^2)$; $Mn^{2+}(3d^5)$, $Mn^{3+}(3d^4)$, $Mn^{4+}(3d^3)$; $Fe^{2+}(3d^6)$, $Fe^{3+}(3d^5)$, $Fe^{4+}(3d^4)$; $Co^{2+}(3d^7)$, $Co^{3+}(3d^6)$; $Ni^{2+}(3d^8)$, $Ni^{3+}(3d^7)$, $Ni^{4+}(3d^6)$; $Cu^+(3d^{10})$, $Cu^{2+}(3d^9)$, $Cu^{3+}(4d^8)$; $Mo^{3+}(4d^3)$, $Mo^{4+}(4d^2)$, $Mo^{5+}(4d^1)$; $W^{4+}(5d^2)$. The most studied TM ion in biology is **iron** [36], which occurs in three major oxidation states and several spin (S) states, namely, (a) $Fe^{2+}(3d^6)$: ferrous - EPR difficult (non-Kramers ion): S = 2 or S = 0; (b) $Fe^{3+}(3d^5)$: ferric - usual EPR probe, the S-state ion: S = 5/2, 3/2, $^1/_2$; (c) $Fe^{4+}(3d^4)$: ferryl - EPR difficult (non-Kramers ion): S = 2 or S = 0. Tables presenting (1) the classification of TM ions in bio-systems together with pertinent references and (2) representative most recent EMR studies of TM centers in biological systems, will be included in the full review. Topics to be covered in the full review include: (i) variety of biological ligands, (ii) classification of protein-bound metal sites by functions [26], (iii) role of EMR in studies of TM bio-systems, and (iv) the specific information provided by EPR on iron centers.

EMR of TM ions in bio-systems is a vast and mature yet still an important and growing area. Major recent reviews dealing with the area in question are: Hagen (1992), Cammack & Cooper (1993), Gaffney & Silverstone (1993), Hanson (1993), Bencini & Zanchini (1993), Basosi *et al.*, (1993), Hoffman *et al.*, (1993), Thomann & Bernardo (1993), Brudvig (1995), DeRose & Hoffman (1995), Hüttermann (1993, 1996), and Cammack & Shergill (1998). Trends in the area of EMR studies of TM ions in biologically important systems can be summarized as follows.

- **Driven by further challenges in research,** for example, shifting attention to protein sites not yet synthesized in laboratory, unlike the family of iron-sulfur redox centers, which has been well studied [26].

- **Significant research opportunities** arising from a plethora of poorly understood microscopic issues in the underlying bio-systems, which include various fundamental properties.
- Aspects in i**ron-sulfur proteins** studies identified by Hagen (1992)—see also Hagen *et al.*, (1998):

  - ■ opening up of the area of superclusters containing Fe-S redox enzymes,
  - ■ important developments in our understanding of high-spin EMR,
  - ■ more studies of the mechanisms of double-exchange interactions,
  - ■ more work on the concept of the g-strain,
  - ■ better theoretical understanding of the mixed spin states systems.

- **Increased applications of the theoretical interpretation** of EMR spectra worked out for the disordered and low symmetry inorganic materials, for example, glasses and minerals, respectively, to studies of proteins, which exhibit virtually no symmetry around paramagnetic centers [17].
- **Increased applications of the multifrequency EMR as well as HMF and/or HF EMR** techniques to studies of paramagnetic centers in bio-systems—see Section on HMF/HF EMR of bio-systems

HMF/HF EMR of bio-systems

The definition of HF-EMR, which is gaining nowadays a wider acceptance, considers as **high** the frequencies above W-band (94 GHz), that is, for the first commercially available cw EPR and ENDOR spectrometer at W-band [53]. Some authors use the frequencies above 80 GHz [6] or 90 GHz [22] as a benchmark. There is not a generally accepted definition of HMF EMR, as the strength of the magnetic field B_0 varies with techniques, which usually combine both HF and HMF aspects. For example, the facilities with capabilities for frequencies up to 700 GHz and B_0 up to 25 T [21] and from 24 GHz to 4 THz and B_0 up to 17 T [22] are available. A working definition of HMF-EMR as EMR at B_0 greater or equal to 3 T may be safely adopted [6, 12].

Note that the dominant technique is so far HMF/HF EPR. There is no established and widely accepted acronyms, for example, the following acronyms have been used in the literature: very high-field/high frequency EMR: VHF-EMR [21], high frequency EPR: HF-EPR [45, 1], and very high-field EPR: VHF-EPR [33, 34]. The instrumentation aspects are beyond the scope of this paper, we only mention here that HMF/HF EPR offers a better signal to noise ratio for many biological samples, however, there are still a lot of technological challenges. Note, for example, that the parallel mode $B_0 \parallel B_1$, which offers unique opportunities for the integer spin systems (non-Kramers ions), requires a special cavity, whereas the high price of the commercial W-band spectrometer prevents routine applications of HMF/HF EPR technique as yet. Topics to be covered in the full review include: (i) an outline of the scientific problems driving HMF/HF EPR spectroscopy (see, [11, 12]), (ii) major advantages of HMF/HF EPR and relevant illustrative applications, and (iii) representative most recent HMF/HF EPR studies of biological systems.

HMF- and HF-EMR techniques and their applications to bio-systems experience a rapid growth in recent years and are becoming a very promising tool. Major recent reviews dealing with the area in question are: Eaton & Eaton (1993), Lebedev (1994a and b), Prisner *et al.*, (1994), M'bius (1993, 1995), Brunel (1996), Reijerse *et al.*, (1998), Barra *et al.*, (1998), Eaton & Eaton (1999a, 1999b—very comprehensive review covering the literature through early 1996), Earle & Freed (1999), Hassan et al (1999). Trends in HMF/HF EMR studies of bio-systems, and the features, which underlie them, can be summarized [11, 12] as follows.

- Due to high sensitivity HMF/HF EMR offers a promise of **microscoping imaging** of materials, which are expected to grow significantly.
- Shorter times offered for spectra recording facilitate access to new processes such as **protein motion and protein folding** and/or misfolding, that is, studies of aspects which are very relevant to many diseases.
- Higher sensitivity for small samples enhances a possibility of **magnetic-resonance-force microscopy** applications.
- **Monitoring interactions** between domains and estimating distances in biological systems between two electron spins as well as between electron spins and nuclear spins becomes feasible—such information is of relevance for the genome projects.
- **Spin-polarized spectra** of coupled radical pairs and triplet probes may become more widely used in biological application of HMF/HF EMR spectroscopy.
- **Resolution of the structure of the photosynthetic reaction centers** may enable location of systems, which cannot be observed by other techniques, for example, the pigments omitted by X-ray measurements.
- **Multi-frequency HMF-EMR** (especially in the high frequency ranges) offers a better understanding of (i) electron spin relaxation, (ii) orientational selectivity, and (iii) energy transfer and reactions.
- **Wider applications of the parallel mode** ($B_0 \parallel B_1$) for the *non-Kramers centres* with S = 1 and 2 and very high integer spin systems (S = 9 & 10) can be envisaged.

EMR of Nitric Oxide (NO) and its Biology

Electronic structure determines the paramagnetic properties of nitric oxide (NO) and hence the applicability of EMR techniques [54]. The molecule NO has 11 valence electrons, but the ground state is 'EPR-silent' and EPR spectra only from excited states of the molecule can be observed. However, the free NO^- anion with 12 electrons is paramagnetic in its ground state and EMR is very useful to study the various important biological functions of NO. Briefly, the EMR studies of NO biology utilize NO spin traps, that is, diamagnetic molecules forming adducts with short-lived reactive free radicals. Usually the adducts are less reactive than free radicals and they can be detected by EMR to reveal the identity of free radicals. Since NO decays in physiological systems within seconds, the identification of the more stable NO products using EMR of spin traps is an important technique and provides a lot of very useful information. Topics to be covered in the full review include: (i) an outline of key issues and the role of EMR spectroscopy see, [54] and (ii) an overview of recent EMR studies of NO and its biology subdivided into two groups: (1) organic spin traps, that is, mostly nitroso or nitrone groups, aimed at elucidation of the mechanism of the free radicals reactions and monitoring NO production, and (2) transition metal complexes.

EMR of NO and its biology is an especially important interdisciplinary and growing area. It is worth mentioning that the Nobel Medicine Price 1998 was awarded to R. Furchgott, F. Murad and L. Ignarro for *Research on How NO Acts on the Human Body*. Major recent reviews dealing with the area in question are: Marsh & Horvath (1989), Tomasi & Iannone (1993), Thomas *et al.*, (1993), Millhauser *et al.*, (1995), Singel & Lancaster (1996), Henry & Singel (1996), Kosaka & Shiga (1996), Korth & Weber (1996), Smirnov *et al.*, (1998); a few most recent papers discussed below are, for example, Bennati et al (1999), Liang & Freed (1999), Park *et al.*, (2000), Strancar *et al.*, (2000). Trends in the EMR studies of NO can be summarized, depending on a given EMR technique, as follows:

- **Conventional EPR:**
  Here the specific topics include, for example, (i) NO serving as a structural probe of the large complex of proteins to study the production, uptake and aspects of NO biology [54] (ii) denitrification process for the NO produced as an atmospheric pollutant arising from exhaust gases (from automobile, aircraft, and industrial fumes)—this may lead to better methods of reduction and/or removal of NO [54, 23] and (iii) characterization of the membranes of biological samples using nitroxides - note the computational methods to extract useful information from EPR spectral lineshapes recently worked out [56].
- **HMF/HF EPR**
  Here the specific topics include, for example, (i) nitroxyl spin probes and labels, which are specially attractive due to the better resolution of g-anisotropy (g_x, g_y, g_z) and the site selectivity; (ii) molecular structure of organic and bioorganic systems containing other free radicals, organic biradicals, paramagnetic ions, and triplet states; and (iii) spin dynamics, molecular dynamics, and chemical kinetics studies [41, 55, 4].
- **Multifrequency EMR**
  Here the specific topics include study of complex dynamics of biomolecules, for example, nitroxide spin-labeled proteins, T4 lysozome, spin-labeled DNA nucleosides.

EMR ands Photosynthesis Research
Photosynthesis, is the conversion of sunlight into chemical energy [42, 29, 14, 15, 37]. In some bacteria, photosynthesis is a simpler process, since bacteria assimilate CO_2, but cannot split H_2O. Photosynthesis occurs in the so-called reaction centers (RC). In green plants, photosynthesis is a more complex process

and occurs either in the so-called photosystems I (P I) or photosystems II (P II) according to the reaction formula:

$$CO_2 + H_2O \xrightarrow[\text{chlorophyll}]{\text{light}(h\nu)} C\,(H_2O) + O_2$$

Briefly, the process of photosynthesis [42, 29, 14, 15, 37] is mediated in both bacteria and plants by the reaction centers, which are integral membrane-protein complexes. Basic 'primary process' in a RC consist of the light-induced charge separation, that is, electron transfer, which can be represented as follows:

$$DX \underset{K_{AD}}{\overset{h\nu}{\square}} D^{+\bullet}\; X^{-\bullet}$$

where D denotes the primary donor, X the primary acceptor, K_{AD} is the charge recombination rate, which represents the reverse ('*wasteful*') process to the electron transfer reactions.

The significant role of EMR in photosynthesis research arises from the fact that in the electron transfer processes, due to the charge separation, two species with unpaired spins, that is, $D^{+\bullet}$ and $X^{-\bullet}$, are produced. Hence EMR techniques, both cw and pulsed modes, are ideal techniques to study, on the molecular level, the primary reactants, that is, transient radical ion intermediates. EMR provides detailed information about, for example, (i) electronic and geometrical structures, and (ii) mobility of the primary reactants. Thus EMR spectroscopy is *complementary* to the X-ray crystallography, which provides directly the s*tructural* information, and to the fast laser spectroscopy, which provides the *kinetic* information. Judging by the vast literature accumulated over the years, it can be stated that EMR spectroscopy, especially the rapidly growing HMF/HF EMR as well as multifrequency EPR and ENDOR, are most powerful tools in photosynthesis research. An overview of major points concerning the EMR studies of photosynthesis subdivided into the four major areas: (i) HMF EPR and ENDOR, (ii) conventional EPR, and (iii) multiple and pulsed EMR (ENDOR and ELDOR), (iv) ODMR of triplet states, will be covered in the full review.

Major recent reviews dealing with the area in question are: Hoff (1993), Möbius (1993, 1995), Maki (1995), Kawamori et al (1995), Feher (1998), Feher & Okamura (1998), Lubitz & Feher (1999). Trends in the EMR studies of photosynthesis can be summarized, depending on a given EMR technique, as follows:

- **Conventional and pulsed EMR**
  Here the specific topics include, for example, (i) the "2+1" ESE study of three dimensional structure of PS II, (ii) time resolved EPR studies of biomimetic model systems, that is, porphyrins covalently linked to quinines (P-Qs), which enable determination of the factors controlling electron transfer in photosystems, (iii) spin polarized transient EPR studies of coupled three-spin systems, which require development of new theoretical methods for description and simulation of EMR spectra, (iv) multiple and pulsed ENDOR as well as ELDOR studies of the structure and function of PS II as well as the role of Mn^{2+} and Fe^{3+}.

- **HMF/HF and multifrequency EPR**
  Significant increase in the applications of these novel EMR techniques to photosynthesis research has been observed. Due especially to the better resolution of **g**-anisotropy (g_x, g_y, g_z) and hyperfine tensor offered by HMF/HF EPR, deeper insight into the process of photosynthesis can be obtained. Current research efforts focus on understanding the unsolved problems, for example, (i) determination of spin density for the quinone radicals $QA^{-\bullet}$ and $QB^{-\bullet}$, (ii) geometry of hydrogen bonded protons (H-bonds), (iii) symmetry of quinone binding sites, (iv) role of Cu^{2+} or Co^{2+} substituted for Fe^{2+} in photosystems, (v) light-induced structural changes (Lubitz & Feher 1999).
- **ODMR of triplet states**
  Extension of ODMR of triplet states to study the metal ligands in metalloproteins has been envisaged.

Acknowledgments

The author is grateful to Prof. K.P. Mishra for his invitation to deliver an invited talk at the 88th Indian Science Congress as well as encouragement and patience. My apology is extended to all colleagues whose work has not been fully commented on due to lack of space. The RGC and the City University of Hong Kong supported this work through the research grant: SRG 7000965 and a conference grant. Technical assistance with the manuscript by Miss Hilda Fung is appreciated.

References

1. Barra, A.L., L.C. Brunel, D. Gatteschi, L. Pardi and R. Sessoli (1998) High-frequency EPR spectroscopy of large metal ion clusters: from zero field splitting to quantum tunneling of the magnetization. *Acc. Chem. Res* 31: 460–466.
2. Basosi, R., W.E. Antholine and J.S. Hyde (1993) Multifrequency ESR of copper biophysical applications *Biological Magnetic Resonance,* vol 13: EMR of Paramagnetic Molecules, eds L.J. Berliner and J. Reuben (New York: Plenum Press) pp. 103–150.
3. Bencini, A. and C. Zanchini (1991) Transition metal ions *A Specialist Periodical Report Electron Spin Resonance (Cambridge: Royal Society of Chemistry)* vol. 12B: pp 1–63.
4. Bennati, M., G.J. Gerfen, G.V. Martinez, D.J. Singel and G.L. Millhause (1999) Nitroxide side-chain dynamics in a spin-labeled helix-forming peptide revealed by high-frequency (139.5-GHz) EPR spectroscopy *J. Magn. Reson.* 139: 281–286.
5. Brudvig, G.W. (1995) Electron paramagnetic resonance spectroscopy *Methods in Enzymology,* Vol 246: Biochemical Spectroscopy, ed K. Sauer (San Diego: Academic Press) pp 536–554.
6. Brunel, L.C. (1996) Recent developments in high frequency/high magnetic field CW EPR applications in Chemistry and biology *Appl. Magn. Reson.* 11: 417–423.
7. Cammack, R. and C.E. Cooper (1993) Electron paramagnetic resonance spectroscopy of iron complexes and iron containing proteins *Methods in Enzymology,*

Vol 227: Metallobiochemistry, eds J F Riordan and B L Vallee (San Diego: Academic Press) pp 353–384.

8. Cammack, R. and J.K. Shergill (1998) Bio-medical applications of EPR spectroscopy *Modern Applications of EPR/ESR from Biophysics to Materials Science* eds C Z Rudowicz, K N Yu and H Hiraoka (Singapore: Springer) pp 3–12.

9. DeRose, V.J. and B.M. Hoffman (1995) Protein structure and mechanism studied by electron nuclear double resonance spectroscopy. *Methods in Enzymology,* Vol 246: Biochemical Spectroscopy, ed K. Sauer (San Diego: Academic Press) pp. 554–589.

10. Earle, K.A. and J.H. Freed (1999) Quasioptical hardware for a flexible FIR-EPR spectrometer *Appl. Magn. Reson.* 16: 247–272.

11. Eaton, G.R. and S.S. Eaton (1999a) High-field and high-frequency EPR *Appl. Magn. Reson.* 16: 161–166.

12. Eaton, G.R. and S.S. Eaton (1999b) ESR imaging *Handbook of Electron Spin Resonance* vol 2 eds C.P. Poole Jr. and H.A. Farach (New York: Springer) pp 327–343.

13. Eaton, S.S. and G.R. Eaton (1993) Applications of high magnetic fields in EPR spectroscopy *Magn. Reson. Rev.* 16: 157–181.

14. Feher, G. (1998) The primary and secondary electron acceptors in bacterial photosynthesis: I. A chronological account of their identification by EPR *Appl. Magn. Reson.* 15: 23–38.

15. Feher, G. and M.Y. Okamura (1999) The primary and secondary acceptors in bacterial photosynthesis: II. The structure of the Fe^{2+}-Q^- complex *Appl. Magn. Reson.* 16: 63–100.

16. Gaffney, B.J. and H. Silverstone (1993) Simulation of the EMR spectra of high-spin iron in proteins *Biological Magnetic Resonance,* vol 13: EMR of Paramagnetic Molecules, eds L.J. Berliner and J Reuben (New York: Plenum Press) pp 1–58.

17. Gaffney, B.J., B.C. Maguire, R.T. Weber and G.G. Maresch (1999) Disorder at metal sites in proteins: a high- frequency-EMR study *Appl. Magn. Reson.* 16: 207–221.

18. Hagen, W.R. (1992) EPR spectroscopy of iron-sulfur proteins *Adv. Inorg. Chem.* 38: 165–215.

19. Hagen, W.R., W.A.M. van den Berg, W.M.A.M. van Dongen, E.J. Reijerse and van Kan (1998) EPR spectroscopy of biological iron-sulfur clusters with spin-admixed S = 3/2 ground states *J. Chem. Soc.-Fara. Tran.* 94: 2969–2973.

20. Hanson, G.R. (1993) Metalloproteins *A Specialist Periodical Report Electron Spin Resonance* (Cambridge: Royal Society of Chemistry) vol 13B: pp 86–130.

21. Hassan, A.K., A.L. Maniero van, H. Tol, C. Saylor and L.C. Brunel (1999) High-field EMR: Recent CW developments at 25 Tesla, and next-millennium challenges *Appl. Magn. Reson.* 16: 299–308.

22. Hassan, A.K., L.A. Pardi, J. Krzystek, A. Sienkiewicz, P. Goy, M. Rohrer and L.C. Brunel (2000) Ultrawide band multifrequency high-field EMR technique: a methodology for increasing spectroscopic information *J. Magn. Reson.* 142: 300–312.

23. Henry, Y.A. and D.J. Singel (1996) Metal-nitrosyl interactions in nitric oxide biology probed by electron paramagnetic resonance spectroscopy *Methods in Nitric Oxide Research*, eds M. Feelisch and J.S. Stamler (Chichester: John Wiley) pp 357–372.

24. Hoff, A.J. (1993) Optically detected magnetic resonance of triplet states in proteins *Methods in Enzymology,* Vol 227: Metallobiochemistry, eds J.F. Riordan and B. L. Vallee (San Diego: Academic Press) pp 290–330.

25. Hoffman, B.M., V.J. DeRose, P.E. Doean, R.J. Gurbiel, A.L.P. Houseman and J. Telser (1993) Metalloenzyme active-site structure and function through multifrequency CW and pulsed ENDOR *Biological Magnetic Resonance,* vol 13: EMR of Paramagnetic Molecules, eds L.J. Berliner and J. Reuben (New York: Plenum Press) pp 151–218.

26. Holm, R.H., P. Kennepohl and E.I. Solomon (1996) Structural and functional aspects of metal sites in biology *Chem. Rev.* 96: 2239–2314.

27. Hüttermann, J. (1993) ENDOR of randomly oriented mononuclear metalloproteins toward structural determinations of the prosthetic group *Biological Magnetic Resonance,* vol 13: EMR of Paramagnetic Molecules, eds L.J. Berliner and J. Reuben (New York: Plenum Press) pp. 219–252.

28. Hüttermann, J. (1996) EPR and ENDOR of metalloproteins *A Specialist Periodical Report Electron Spin Resonance* vol 15: (Cambridge: Royal Society of Chemistry) pp. 59–111.

29. Kawamori, A., H. Mino, H. Hara, A.V. Astashkin and Y.D. Tsvetkov (1995) Multiple resonance in pulsed EPR with application to photosynthesis *Annual Studies* vol XLIV (Nishinomiya: Kwansei Gakuin University) pp. 221–239.

30. Korth, H.G. and H. Weber (1996) Detection of nitric oxide with nitric oxide-trapping reagents *Methods in Nitric Oxide Research,* eds M. Feelisch and J.S. Stamler (Chichester: John Wiley) pp. 383–392.

31. Kosaka, H. and T. Shiga (1996) Detection of nitric oxide by electron spin resonance using hemoglobin *Methods in Nitric Oxide Research,* eds M. Feelisch and J.S. Stamler (Chichester: John Wiley) pp. 373–381.

32. Kurreck, H., G. Elger, J. von Gersdorff, A. Wiehe and K. Möbius (1998) EPR studies of photoinduced electron transfer in triad model compounds of photosynthesis *Appl. Magn. Reson.* 14: 203–215.

33. Lebedev, Y.S. (1994a) High-field ESR *A Specialist Periodical Report Electron Spin Resonance* vol 14: (Cambridge: Royal Society of Chemistry) pp. 63–87.

34. Lebedev, Y.S. (1994b) Very-high-field EPR and its applications *Appl. Magn. Reson.* 7: 339–362.

35. Liang, Z. and J.H. Freed (1999) An assessment of the applicability of multifrequency ESR to study the complex Dynamics of biomolecules *J. Phys. Chem. B* 103: 6384–6396.

36. Lindley, P.F. (1996) Iron in biology: a structural viewpoing. *Rep. Prog. Phys.* 59: 867–933.

37. Lubitz, W. and G. Feher (1999) The primary and secondary acceptors in bacterial photosynthesis III. Characterization of the quinone radicals Q_A^- and Q_B^- by EPR an ENDOR *Appl. Magn. Reson.* 17: 1–48.

38. Maki, A.H. (1995) Optically detected magnetic resonance of photoexcited triplet states *Methods in Enzymology,* Vol. 246: Biochemical Spectroscopy, ed K Sauer (San Diego: Academic Press) pp. 610–638.

39. Marsh, D. and L.I. Horvath (1989) Spin label studies of the structure and dynamics of lipids and proteins in membranes *Advanced EPR - Applications in Biology and Biochemistry,* ed A.J. Hoff (Amsterdam: Elsevier) pp. 707–754.

40. Millhauser, G.L., W.R. Fiori and S.M. Miick (1995) Electron spin labels *Methods in Enzymology,* Vol. 246: Biochemical Spectroscopy, ed K. Sauer (San Diego: Academic Press) pp. 589–610.

41. Möbius, K. (1993) High-field EPR and ENDOR on bioorganic systems *Biological Magnetic Resonance,* Vol 13: EMR of Paramagnetic Molecules, eds L.J. Berliner and J. Reuben (New York: Plenum Press) pp. 253–274.

42. Möbius, K. (1995) High-field/high-frequency EPR/ENDOR – a powerful new tool in photosynthesis research *Appl. Magn. Reson.* 9: 389–407.

43. Park, S.K., V. Kurshev, C.W. Lee and L. Kevan (2000) Electron spin resonance and optical spectroscopic studies of Co-ZSM-5 with nitric oxide *Appl. Magn. Reson.* 19: 21–33.

44. Prisner, T.F., M. Rohrer and K. Mobius (1994) Pulsed 95 GHz high-field EPR heterodyne spectrometer with high spectral and time resolution *Appl. Magn. Reson.* 7: 167–183.

45. Reijerse, E.J., P.J. van Dam, A.A.K. Klaassen, W.R. Hagen, P.J.M. van Bentum and G.M. Smith (1998) Concepts in High-Frequency EPR – Applications to bio-inorganic systems *Appl. Magn. Reson.* 14: 153–167.

46. Rudowicz, C. (1987) Concept of spin Hamiltonian, forms of zero-field splitting and electronic Zeeman Hamiltonians and relations between parameters used in EPR. A critical review *Magn. Res. Rev.* 13: 1–89; *Erratum* 1988 - *ibidem* 13: 335.

47. Rudowicz, C. (1997) Programme and Abstracts Book - the First Asia-Pacific EPR/ESR Symposium City University 20-24 January 1997 (Hong Kong: City University) pp. 1–182.

48. Rudowicz, C. (2001) Trends in Applications of EMR Spectroscopy in Biology, Life and Medical Sciences (*Invited Talk*), 88th Indian Science Congress, Section: Biochemistry, Biophysics & Molecular Biology, Symposium: Emerging Trends in Biomedical Research and Technologies, New Delhi, January 3–7, 2001.

49. Rudowicz, C. (Editor), K.N. Yu and H. Hiraoka (Assoc. Eds) (1998) Modern Applications of EPR/ESR: From Biophysics to materials Science - Proceedings of the First Asia-Pacific EPR/ESR Symposium Hong Kong 20–24 January 1997 (Singapore: Springer) pp. 1–666.

50. Rudowicz, C. and S.K. Misra (2001) Spin-Hamiltonian Formalisms in Electron Magnetic Resonance (EMR) & Related Spectroscopies *Appl. Spectr. Rev.* - in press.

51. Rudowicz, C., A. Galeev and T.C. Y. Chung (2001) Operator and Parameter Notations in Spin Hamiltonians used in Electron Magnetic Resonance (EMR) and related spectroscopies *Appl. Spectr. Rev.* - in preparation.

52. Salikhov, K.M., A.J. van der Est and D. Stehlik (1999) The transient EPR spectra and spin dynamics of coupled three-spin systems in photosynthetic reaction centers *Appl. Magn. Reson.* 16: 101–134.

53. Schmalbein, D., G.G. Maresch, A. Kamlowski and P. Hofer (1999) The Bruker high-frequency-EPR system *Appl. Magn. Reson.* 16: 185–205.

54. Singel, D.J. and Jr J.R. Lancaster (1996) Electron paramagnetic resonance spectroscopy and nitric oxide biology *Methods in Nitric Oxide Research*, eds M. Feelisch and J.S. Stamler (Chichester: John WileLancaster Lancaster Lancaster Lancaster Lancaster Lancaster y) pp. 341–356.

55. Smirnov, A.I., R.L. Belford and R.B. Clarkson, (1998) Comparative spin label spectra at X-band and W-band *Biological Magnetic Resonance,* vol 14: Spin Labeling: The next Millennium, ed L J Berliner (New York: Plenum Press) pp. 83–107.

56. Strancar, J., M. Sentjure and M. Schara (2000) Fast and accurate characterization of biological membranes by EPR spectral simulations of nitroxides *J. Magn. Reson.* 142: 254–265.

57. Thomann, H. and M. Bernardo (1993) Pulsed electron nuclear double and multiple resonance spectroscopy of metals in proteins and enzymes *Biological Magnetic Resonance,* vol. 13: EMR of Paramagnetic Molecules, eds L.J. Berliner and J. Reuben (New York: Plenum Press) pp. 275–322.

58. Thomas, D.D., M. Ostap, C.L. Berger, S.M. Lewis, P.G. Fajer and J.E. Mahaney (1993) Transient EPR of spin- labeled proteins *Biological Magnetic Resonance,* vol 13: EMR of Paramagnetic Molecules, eds L.J. Berliner and J. Reuben (New York: Plenum Press) pp. 323–351.

59. Tomasi, A. and A. Iannone (1993) ESR spin-trapping artifacts in biological model systems *Biological Magnetic Resonance,* vol 13: EMR of Paramagnetic Molecules, eds L.J. Berliner and J. Reuben (New York: Plenum Press) pp. 353–384.

60. Tonaka, M., A. Kawamori, H. Hara and A.V. Astashkin (2000) Three-dimensional structure of electron transfer components in photosystem II: "2+1" ESE of chlorophyll Z and tyrosine D *Appl. Magn. Reson.* 19: 141–150.

61. Xu, Y. (1999) Programme and Abstracts Book - the Second Asia-Pacific EPR/ESR Symposium, Zejiang University Hangzhou P.R. China. October 31-November 4 Hangzhou: Zejiang University pp. 1–184.

Radiobiology and Bio-Medical Research
Edited by K.P. Mishra
Copyright © 2004 Narosa Publishing House, New Delhi, India

24. High Field ESR Measurements of Organic Compounds

H. Ohta[1], S. Okubo[1], T. Sakurai[1], K. Kirita[1], K. Kanoda[2],
K. Hiraki[3], K. Enomoto[4], A. Miyazaki[4], T. Enoki[4],
H. Yamamoto[5], R. Kato[5]

[1]Department of Physics and Venture Business Laboratory, Kobe University,
1-1 Rokkodai, Nada, Kobe 657-8501, Japan

[2]Department of Applied Physics, University of Tokyo, 7-3-1 Hongo,
Bunkyo-ku, Tokyo 113-8656, Japan

[3]Department of Physics, Gakushuin University, 1-5-1 Mejiro, Toshima-ku,
Tokyo 171-8588, Japan

[4]Department of Chemistry, Tokyo Institute of Technology, Ookayama,
Meguro-ku, Tokyo 152-8551, Japan

[5]Institute of Physical and Chemical Research, Wako, Saitama 350-0198, Japan

Abstract: High field electron spin resonance (ESR) measurements using
high frequency and high magnetic field has an advantage of the high resolution
compared to a conventional X-band ESR, and it is promising for the study
of the biochemical substances. High field ESR system in Kobe covers the
frequency region from 30 to 1,200 GHz and the magnetic field region up to
30 T by the pulsed magnetic field. The temperature dependence measurement
can be performed from 1.8 K to 300 K. High field ESR results of several
organic compounds, $(DI\text{-}DCNQI)_2Ag$, $(DMET)_2FeBr_4$ and $Et_2Me_2P[Pd
(dmit)_2]_2$, are presented and discussed.

Introduction

Electron spin resonance (ESR) measurement is a powerful means to study the
electronic states of radicals or magnetic materials and it is used in a wide scientific
field such as physics, chemistry, biology and medicine. Usually ESR measurement
is performed by a conventional ESR system equipped with X-band (about 9
GHz) source and water cooling magnet up to 1 T to fulfil the resonance condition

$$h\nu = g\mu_B H \tag{1}$$

where h, ν, g, μ_B and H are the Plank constant, the frequency, the g-value, Bohr
magneton and the external magnetic field, respectively. The g-value will give the
information about the electronic states. On the other hand, the high field ESR
uses much higher frequency than X-band source. Therefore, the high field ESR
requires much higher magnetic field to fulfil the resonance condition (1), and it

uses superconducting magnet or pulsed magnet which is not so easy to access for the scientists in the field except those from the field of physics. However, the high field ESR has the following advantages compared to the conventional X-band ESR [10, 8].

(1) High resolution.
(2) Strong absorption intensity.
(3) Very broad absorption line can be detected.
(4) Antiferromagnetic resonance (AFMR) with large AFMR gap or direct transition between the spin gap can be detected.
(5) ESR beyond the magnetic phase transition can be detected.

(The advantage 1) is especially important for the investigation in the field of biology or medicine, and the application of the high field ESR to these fields is the final goal of its development. However, the increase of the sensitivity is especially required. In this paper the development of high field ESR in Kobe is presented by showing some applications to the organic magnetic materials.

High Field ESR System in Kobe

Available light sources in Kobe University are the following.

(a) Gunn oscillators 30–160 GHz
(b) Multipliers of Gunn oscillators 180, 210, 220, 240, 300, 315 GHz
(c) Backward traveling wave oscillators (BWO) 117.7–1183.6 GHz

Using these light sources, we can cover the millimeter and submillimeter wave region as shown in Fig. 1. For the high field, the pulsed magnetic field is used. $H_{max} = 30$ T is produced by the discharge of the 23.5 kJ capacitor bank ($V_{max} = 3$ kV) into the ice magnet immersed in the liquid nitrogen. Recently we introduced a new 100 kJ capacitor bank ($V_{max} = 3$ kV) and succeeded in producing 42 T by a pulsed magnet using Cu–Ag wire [3]. The transmitted light through the sample, which is situated at the center of the magnet, is detected by the liquid we cooled

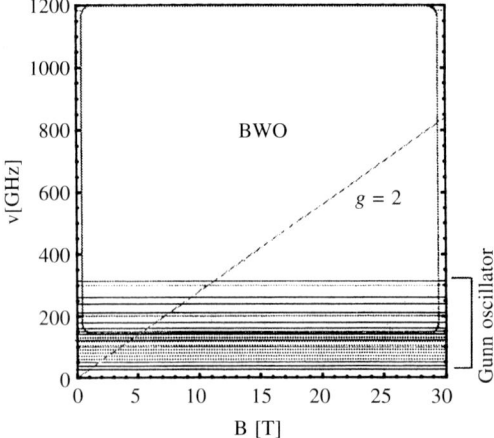

Figure 1. Available frequencies and magnetic field at Kobe University.

InSb detector. The sample temperature can be varied from 1.8 to 300 K. The signals from the detector and the pick up coil, which is monitoring the magnetic field, during the 7 msec pulsed magnetic field are stored in the two channels digital memory. The details of our high field ESR system can be found in refs [8–11].

Antiferromagnetic Resonance Measurements (AFMR) of (DI–DCNQI)$_2$Ag

As the size of the organic single crystal is very small, for instance $0.1 \times 0.1 \times 1$ mm^3, compared to the inorganic single crystals, the determination of the magnetic anisotropy by conventional magnetic susceptibility or magnetization measurements by SQUID is difficult. Moreover, the existence of the large dimagnetism in the organic substance makes the measurement more difficult. However, the observation of AFMR by high field ESR enables the determination of the magnetic anisotropy. Although DI–DCNQI molecules form one dimensional columns along c-axis in (DI–DCNQI)$_2$Ag, it undergoes an antiferromagnetic (AF) order below $T_N = 5.5$ K due to the interchain interactions. Although the AF order was difficult to be detected by SQUID measurement, it was confirmed by the 1H-NMR measurement [5] and AFMR measurement by our high field ESR [7]. Figure 2 shows the frequency-field relation of observed AFMR at 1.8 K. The frequency-field relation coincides well with the conventional AFMR theory and the results clearly show that the c-axis is the easy axis. As the AFMR shifts from electron paramagnetic resonance (EPR) due to the formation of the internal field below T_N, we can estimate the magnitude of the magnetic anisotropy from the zero field AFMR gap of about 1.2 T. From the AFMR analysis, the anisotropic field is estimated to be 3 mT, which is comparable with the other organic antiferromagnet b'-

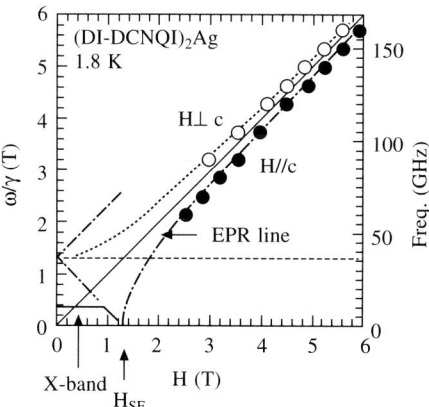

Figure 2. Frequency-field relation of AFMR in (DI-DCNQI)$_2$Ag observed at 1.8 K. Solid and open circles correspond to H//c and H⊥c, respectively. The solid line shows the observation region of a conventional X-band ESR. Dashed-dotted and dotted lines are AFMR modes calculated by a conventional molecular field AFMR theory.

(BEDT-TTF)$_2$ICl$_2$ [3]. We have to point out that the AFMR cannot be detected by a conventional X-band ESR shown by the solid line in Figure 2, and the use of the high field ESR is essential for the observation of AFMR in the case of (DI-DCNQI)$_2$Ag.

High Field ESR Measurement of p-d Electrons System (DMET)$_2$FeBr$_4$

(DMET)$_2$FeBr$_4$ is an organic conductor whose structure consists of an alternating stacking of 1 dimensional chain based on DMET donar molecules and magnetic Fe^{3+} (S = 5/2) square lattices. The system is metallic due to π-electrons and shows metal-insulator (MI) transition at T_{MI} = 40K due to the formation of spin density wave (SDW), while Fe^{3+} spins undergo an antiferromagnetic transition at T_N = 3.7 K [4]. A strong correlation between the magnetization processes and the magnetoresistances at low temperature suggests the presence of strong π-d interactions. As the line width of Fe^{3+} ESR is very broad, the high field ESR has the advantage to observe the resonance and to study the frequency dependence. The aim of our study is to see the effect of strong π-d interactions in (DMET)$_2$FeBr$_4$ by observing ESR by our high field ESR.

Figure 3 shows the temperature dependence of the resonance field of observed ESR. The inset shows the typical transmission spectra. It is reasonable to consider that the sharp absorption line is from π electron and the broader absorption line is from Fe^{3+}. We have to point out that the observation of both ESR lines became possible due to the fact that we used our high field ESR system. We can clearly see the splitting of two resonances below T_{MI} suggesting the existence of the internal field due to the formation of SDW below T_{MI}. Such shift of resonances in π-d interaction system is observed by our high field ESR for the first time. To

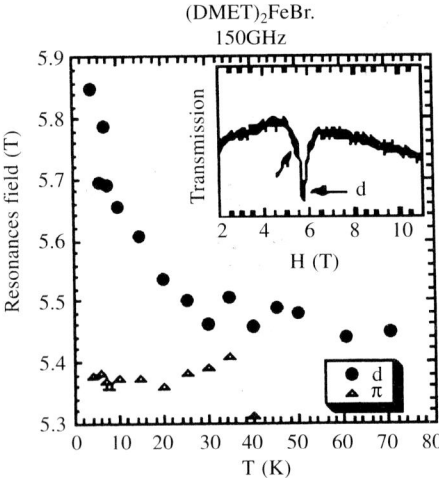

Figure 3. Temperature dependence of resonance fields of π and d electrons observed at 150 GHz. Powder samples of (DMET)$_2$FeBr$_4$ are used. The inset shows the typical transmission spectra at 8 K.

clarify the SDW state below T_{MI}, detailed frequency dependence measurements are required.

ESR Measurements of $Et_2Me_2P[Pd(dmit)_2]_2$

$Pd(dmit)_2$ compounds have the one dimensional LUMO band located below the two dimensional HOMO band and show rich varieties of physical properties [10]. For instance, β-$Me_4N[Pd(dmit)_2]_2$ is a superconductor under high pressure ($T_c = 6.2$ K at 6.5 kbar [9]) while there are many Mott-Hubbard insulators which undergo antiferromagnetic order. We are starting systematic studies of the system by ESR measurements [2]. Here our recent ESR results on $Et_2Me_2P[Pd(dmit)_2]_2$ will be shown.

$Et_2Me_2P[Pd(dmit)_2]_2$ shows antiferromagnetic order below 18 K which is suggested by X-band ESR measurement [12]. It loses its integrated intensity rapidly at 18 K as shown in Figure 4. It also decreases at 18 K for H// a and b axes. Therefore the onset of antiferromagnetic order is suggested at 18 K. The line width also decreases as the temperature decreases. However, the situation is completely different for the high field ESR. The integrated intensity and line width tend to increase as the temperature is decreased at 140 GHz as shown in Figure 5. Figure 6 shows the temperature dependence of g-values observed at different frequencies. It is clear that the g-shifts start at around 18 K suggesting the existence of phase transition. However, the temperature dependence between X-band (9.43 GHz) and other frequencies above 140 GHz is completely different and the detailed frequency dependence measurement will be the key to understand the origin of the phase transition at 18 K in $Et_2Me_2P[Pd(dmit)_2]_2$.

Summary

High field ESR system at Kobe University is introduced and the application of our high field ESR to organic compounds, $(DI-DCNQI)_2Ag$, $(DMET)_2FeBr_4$ and $Et_2Me_2P[Pd(dmit)_2]_2$ are presented. Comparison with the conventional X-band ESR is made, and we showed the advantages and complimental information from the high field ESR measurements.

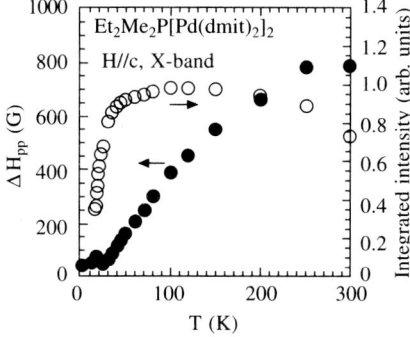

Figure 4. **Temperature dependence of line width and integrated intensity of $Et_2Me_2P[Pd(dmit)_2]_2$ in observed by X-band ESR.**

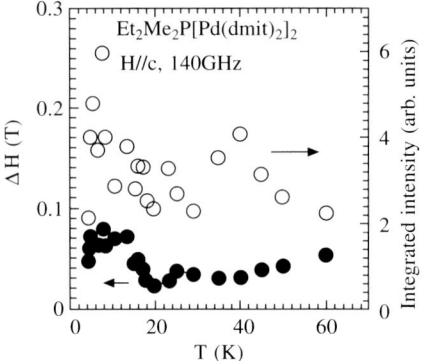

Figure 5. Temperature dependence of line width and integrated intensity of Et₂Me₂P[Pd(dmit)₂]₂ observed at 140 GHz.

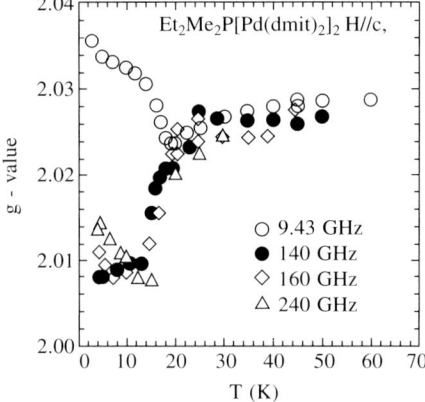

Figure 6. Temperature dependence of g-values observed for H//c.

Acknowledgments

This work was partly supported by a Grants-in-Aid for scientific research on priority Area (No. 11136231'Metal-assembled Complexes') from the Ministry of Education, Science, Sports and Culture of Japan.

References

1. Ohta, H., S. Okubo, T. Sakurai, T. Goto, K. Kirita, K. Ueda, Y. Uwatoko, T. Saito, M. Azuma, M. Takano and J. Akimitsu. To appear in Physica B
2. Ohta, H., T. Sakurai, S. Okubo, R. Kato and T. Nakamura, to appear in Synth. Metal.
3. Akioka, K. H. Ohta, S. Okubo, K. Miyagawa and K. Kanoda (1998) *J. Magn. Magn. Mater.* 177: 746.
4. Enomoto, K., A. Miyazaki and T. Enoki, to appear in Synth. Metal.
5. Hiraki, K. and K. Kanoda (1996) *Phys. Rev. B* 54: R17276.

6. Motokawa, M., H. Ohta and N. Makita (1991) *Int. J. Infrared & MMW* 12(2): 149.

7. Nakagawa, N., K. Akioka, H. Ohta, S. Okubo, K. Kanoda and K. Hiraki (1999) *Synth. Metal* 103: 1894.

8. Nakagawa, N., T. Yamada, K. Akioka, S. Okubo, S. Kimura and H. Ohta (1998) *Int. J. Infrared & MMW* 19(2): 167.

9. Kato, R., Y-L. Liu, Y. Hosokoshi, and S. Aonuma (1997) *Mol. Cryst. Liq. Cryst.* 296: 217.

10. Eaton, S.S. and G.R. Eaton (1999) In *Handbook of Electron Spin Resonanc.* Vol. 2, eds. C.P. Poole, J.R., and H.A. Farach New York: Springer-Verlag. 345.

11. Kimura, S., H. Ohta, M. Motokawa, S. Mitsudo, W-J Jang, M. Hasegawa and H. Takei (1996) *Int. J. Infrared & MMW* 17(5): 833.

12. Nakamura, T. private communication.

Radiobiology and Bio-Medical Research
Edited by K.P. Mishra
Copyright © 2004 Narosa Publishing House, New Delhi, India

25. Peptides Aggregation and Conformation Phenomena as Studied by Advanced ESR Methods

Yu. D. Tsvetkov

Institute of Chemical Kinetics and Combustion Russian Academy of Sciences,
Siberian Branch, Novosisbisrsk-90, 630090, Russia

Among the membrane active peptides the trichogin peptides have particularly specific properties. This peptide antibiotic

Nt–Aib–Gly–Leu–Aib–Gly–Gly–Leu–Aib–Gly–Ile–Leu–Ct

is composed of eleven amino acids and different N and C terminal groups possesses potent membrane modifying properties. It is of current interest because of its ability to induce membrane leakage *in vivo* and *in vitro* and its potential use as a therapeutic. The peptide's mechanism of action is unknown. Its short length does not allow it to span lipid membranes and thus it cannot form conventional multimeric ion channels as has been suggested for longer peptides with similar sequences. To make an advancement in the field it is imperative that we determine the peptide's structure and aggregation state.

The trichogin peptide contains Aib residues and it was recently demonstrated that these Aib's can be replaced with spin active TOAC residues without loss of function. TOAC has a nitroxide side chain and thus allows for electron spin resonance (ESR) to be used as a powerful and novel structural tool. In this report we summarize some of the results obtained in this field by Russian-Dutch-Italian team during realization of the joint project in the last several years.

The ESR spectra contains information on magnetic dipole-dipole interactions which gives the knowledge on distance between spins or pair space distribution function. Different types of ESR methods can be used in order to obtain this distance information for different regions of molecular distances. In the case when dipole-dipole width is much less than the inhomogeneous lines broadening (distances more then $10 - 15$ Å) the only way to get it is to apply pulsed ESR methods. One of most informative is pulsed electron-electron double resonance (PELDOR).

PELDOR method provides us opportunity for the most direct measurement of weak dipole-dipole couplings. The main advantage of the method is much wider range of dipole couplings measurements compare with the other pulsed ESR methods. The theory and applications to different model polyoriented spin systems will be discussed. The examples of distance measurements for either one only in the range of 15–130 Å will be given for the different simple systems together with the determination of numbers of spins in spin aggregates, which contains fixed number of radicals [1]. The peculiarities of phase relaxation and the methods of distance measurements for the radicals pairs will be discussed. It was shown that for three dimensional uniformly distributed neutral radicals the kinetics of spin phase relaxation measured by PELDOR is described by exponential law. But the phase relaxation kinetics deviates from the exponential one in the case of charged radicals. It was shown that electrostatic repulsion effect for the charged radical ions have a satisfactory description based on the Debye-Huckel theory [2]. Spin labeled polymers provide the model systems in which the physical properties depend upon the space structure of polymer chains. In the case of spin labeled poly-4- vinipyridine the intermolecular and intramolecular contribution of dipole-dipole interaction in phase relaxation process have been separated and one dimensional labels pair space distribution function estimated [3].

Using developed theoretical and experimental approach spin labeled peptide trichogin has been extensively studied in glassy solutions by PELDOR. From the modulation patterns which were found in PELDOR signal decay for double labeled peptides the distance between labels (15–20 Å) determined and possible conformation (2_7 or 3_{10}) of helices chain were assumed. It was found that the particular conformation and distance between labels strongly depends upon the structure of terminal groups in peptides and the physical properties of glassy solution [4, 5].

The magnetic dipole-dipole relaxation of mono spin labeled peptides were studied by PELDOR in glassy polar and apolar solvents at 77K. It was established that specific assemblies (aggregates) of trichogin molecules are formed in an apolar solvent but addition of a more polar solvent leads to dissociation of the aggregates [6]. Some of the distances between spin labels in the aggregates have been determined. It is shown that the formation of aggregates in weakly polar solvents is typical of the trichogin analogs. The aggregate structure depends on solvent composition and the structure of terminal groups of peptides. It has been established that the distances that can be measured between spin labels in aggregate groups around the values of 23, 26 and 33 Å. The lower boundary has been experimentally obtained for the mean number of peptide molecules in aggregate in the range of 3.1–4.3 molecules depending on peptide structure and solvent composition. In addition, continuous wave-ESR spectra data suggest the occurrence of aggregated species in the same solutions at room temperature.

The experimental results are consistent with a model wherein four amphiphilic helical peptide molecules form a vesicular system with the polar amino acid side chains pointing to the interior and the apolar side chains to the exterior of the cluster [6, 7].

The PELDOR technique in combination with the continuous wave-ESR methods has been used to investigate the secondary structure of a double spin-labeled peptide that is hidden in a tetrameric supramolecular assembly of unlabeled peptide molecules [8]. The PELDOR signal decay kinetics of spin labels has been experimentally studied in glassy solutions of the double labelled peptide frozen to 77 K in a mixture of chloroform-toluene with an excess of unlabeled peptide. The signal oscillations have been observed at high degrees of dilution with unlabeled peptide. The intramolecular distance between the spin labels of the peptide molecule in the aggregate has been determined from oscillation frequency to be 15.7 Å which is close to the value of ~ 14 Å calculated for a 3_{10}-helical structure. Estimation of the fraction of this ordered secondary structure shows that about 15% of the peptide molecules in aggregates are folded in the 3_{10}-helical conformation.

The author is grateful to Dr. A.D. Milov, Dr. A.G. Maryasov (Novosibirsk), Dr. J. Raap (Leiden), Prof. C. Toniolo, Dr. F. Formaggio, M. Crisma (Padova) who take part in different parts and aspects of this review paper. The work described in this publication was supported in part by The Netherlands Organization of Scientific Research (NWO), project 047.006.009, by the Russian Basic Research Foundation (grants 99–03–33149, 00–15–97321), by the NATO Linkage grant 97194, and by the Award RC1-2056 of the U.S. Civilian Research and Development States of the Former Soviet Union (CRDF).

References

1. Milov, A.D., A.G. Maryasov, Yu.D. Tsvetkov (1998) Appl. Magnetic Reson., 15, 107.
2. Milov, A.D. Yu.D. Tsvetkov. *Appl. Magnetic Reson.* (submitted).
3. Milov, A.D., Yu.D. Tsvetkov (1997) Appl. Magnetic Reson., 12: 495.
4. Milov, A.D., A.G. Maryasov, Yu.D. Tsvetkov (1999) *J. Raap. Chem. Phys. Lett.* 303, 135.
5. Milov A.D., A.G. Maryasov, R.I. Samoilova, Yu.D. Tsvetkov, J. Raap, V. Monaco, F. Formaggio, M. Crisma, C. Toniolo. (2000) *Doklady Acad. Nauk* (Russ), 370: 265.
6. Milov, A.D., D. Yu. F. Tsvetkov, Formaggio, M. Crisma, C. Toniolo. J. Raap, (2000) *J. Am. Chem. Soc.* 122: 3843.
7. Milov, A.D., Yu. D. Tsvetkov, J. Raap. *Appl. Magnetic Reson.* (submited).
8. Milov, A.D. Yu. D. Tsvetkov, F. Formaggio, M. Crisma, C. Toniolo. J. Raap, J. Am. Chem. Soc. (submitted).

Radiobiology and Bio-Medical Research
Edited by K.P. Mishra
Copyright © 2004 Narosa Publishing House, New Delhi, India

26. New Dimensions of Nuclear Techniques in Bio-Medical Sciences

K.K. Dwivedi

Vice-Chancellor, Arunachal University, Rono Hills, Itanagar 791 111, India

Abstract   The nuclear track technique has shown a number of interesting applications in diverse fields including bio-medical sciences due to its potential in producing several application and diagnostic devices. The passage of heavy ions creates nanometric damage trails in polymeric solids which can be revealed as nuclear tracks on chemical etching. The nuclear track technique involves creation, development and manipulation of track (pore) parameters such as shape, size and number density. In this presentation we would like to highlight some of the promising applications of these techniques in bio-medical sciences. These includes (i) special nuclear track micro-filters for treatment of blood cancer and in transdermal therapeutical system (TTS-patches) as control membrane, (ii) the single pore membranes (SPM) as diagnostic tools for carcinoma and for measuring plasticity of blood cells, (iii) development of biocompatible polymer surfaces for cell growth and (iv) applications in cellular radiation biology.

Introduction

The path of an energetic heavy ion through any dielectric material, like a polymer, is a highly damaged zone. Such nanometric sized intense damage trials can be revealed as a 'track' on chemical etching, which can easily be viewed under an optical microscope. The damage trails of the heavy ions, which are revealed, are known as etched tracks and the unetched ones are known as latent tracks. The nuclear track technique utilizes both etched and latent tracks for a wide variety of applications ranging from space physics to oil prospecting and earthquake prediction and from archaeology to bio-medical sciences.

The wide range of applicability of the track technique is an outcome of the versatility of the track characteristics, which can be modified according to need. The term track characteristic incorporates factors such as the track shape, size, length, number density and angle etc. The scope of nuclear track technique in bio-medical sciences [7] is rapidly growing and a number of new innovative applications have come up in this field. A few of these applications require the development and characterization of nuclear track micro-filter (NTMF) while others require single pore membranes (SPM). As the name suggests single pore membranes are composed of a single pore in the film, while the nuclear track micro-filter consists of a number of non-overlapping pores of definite size and shape. Special nuclear track micro-filters are used for the treatment of blood cancer and even in a new method for drug delivery—the transdermal therapeutical system (TTS) patches. Single pore membranes are used as diagnostic tools in cytological applications particularly in determining the plasticity of the blood cells. The SPM can also be used for counting and sizing of bacteria and viruses. Besides these the nuclear track technique has been used in cellular radiation biology to measure cell thickness and also to study the site of damage and the survival rates of the heavy ion irradiated cells [5, 6]. A new method, which is still at an experimental stage is the modification of the polymers to produce bio-compatible surface for cell growth [9]. This would provide the cosmetic and pharmaceutical industries respite from animal activists.

Here, we discuss some of these applications of nuclear track technique in bio-medical sciences.

Materials and Methods

Materials used
Any dielectric solid can be used to develop NTMFs and SPMs. These include organic polymers, namely, polycarbonates (Lexan, Makrofols) polyimides (Kapton), polyethylene Terephthlate (Mylar, Hostaphan), polypropylene, cellulose acetates and cellulose nitrates; minerals such as mica and inorganic solids namely glass. One may select a thickness between a few micrometers to several hundreds of micrometers. The choice of detector material for preparing micro-filters is based on their response to chemical etching in terms of the homogeneity and smoothness of the pore walls after development. The mechanical strength, chemical and radiation resistance of the material also play a great role. The most widely used materials are polycarbonates and polyimides. Recently, thick micro-filters of chemically resistant polypropylene were also developed [11].

Heavy ion irradiation
For producing NTMF, the selected polymer films are cut into circular disks of desired size (diameter = 47 mm or 50 mm) and then mounted onto special Perspex holders for heavy ion irradiation. The thin filters (thickness < 10 μm) were prepared by irradiating films using fission fragments from a ^{252}Cf source while thicker filters (>10 μm) require high-energy heavy ions from accelerators. Generally, a well collimated, diffused and homogenous heavy ion beam is used for irradiation. The films are exposed perpendicular to the surface by heavy ions whose fluence can be precisely controlled as per requirement. Possibilities for large-scale on-line production of NTMFs has also been successfully explored [10, 17]. The single pore membranes (SPM) were prepared by exposing polymeric films to energetic heavy ions in such a way that only one ion goes across the film in each frame. For this the film was continuously moved through a motor and the heavy ion beam was chopped through an electronic chopper to allow only one ion to impinge on the film at a time. After each irradiation the film was advanced by 19 mm for the next exposure. More details on the production of SPM are available elsewhere [13].

Chemical Etching and Characterization

The irradiated films are chemically etched in suitable etchants (NaOH) under optimum conditions. The heavy ion induced damaged trails in polymer films are enlarged at a certain rate in the etchant and the process may be terminated after the desired pore size is achieved. After etching the films are washed carefully to remove etching solution. These are then dried and stored in dust-free environment.

The characterization of NTMFs is carried out in terms of pore-density, inner and outer pore diameters, length of the pores, tapering angle, porosity and throughput. These filter parameters are experimentally determined by using optical and electron microscopes. The pore shape and size are dependent on the nature of the bombarding ion (its mass, charge and energy), the irradiation geometry and etching parameters like relative magnitude of bulk- and track-etch rates. By regulating the etching conditions, one may develop well-defined pores of a desired size and shape. Cylindrical channels are developed when track-etch rate (V_T) is very large as compared to bulk-etch rate (V_G), while tapered (conical) holes are produced by careful adjustment of etching conditions so that the V_T/V_G ratio is not too large.

Applications of Track Devices

The development, characterization and applications of track devices such as NTMF and SPM have been studied extensively [5–13]. Here, we describe a few applications of NTMF and SPM in bio-medical sciences.

(a) Applications in cancer therapy

Nuclear track micro-filters have been put to many diverse applications. One of the early use of NTMF was to separate circulating cancer cells from blood [15] on the basis of a large irregular size and more rigidity of malignant cell as compared to the normal blood cells. The NTMFs can be used to separate and identify cancer cells. Fig.1 shows a NTMF surface on which cancer cells (dark irregular shaped) got stuck after filtration of the blood. Song *et. al.* [16] have

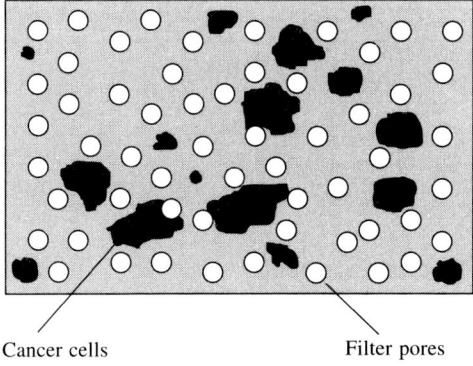

Cancer cells   Filter pores

Figure 1. A schematic diagram showing surface of a NTMF after filtration of blood containing cancer cells (irregular spots).

examined blood samples from patients of breast, colon or rectum cancers and found that the cancer cells were detectable before the malignancy could be metastasized. The cancer cells in spinal fluid and lung materials can also be detected. The NTMFs can also be used in immunology [12] and in the study of metabolic interactions [1] as the membrane partition provides a barrier to cell migration but allows free flow of metabolites, antibodies and antigens.

(b) Applications as transdermal therapeutical system (TTS) patches
The latest application of NTMFs has been in the field of clinical practice. A fairly recent discovery, the Transdermal Therapeutical System (TTS) of drug delivery has substantial advantages over the traditional oral and parenteral routes. This method offers a uniform and constant rate of delivery of a drug without a peak or valley of the traditional methods. Any untoward side effects in the first pass metabolism process of the liver are thus completely eliminated. It also facilitates easy cessation by removal of the patch and, therefore, offers increased patient compliance.

The essential features of a TTS or patch include a drug reservoir and a control membrane to regulate the rate of drug delivery to the surface of the skin. Fig. 2 shows a typical diagram of TTS patch containing various components including NTMF based control system. The control membrane is the key component of the TTS for controlling and sustaining the delivery of drugs at a predictable or optimized rate for therapy. The quantity of drug release per unit area of the membrane can be adjusted over a wide range of convenience simply by increasing the fluence of heavy ions to produce NTMF of high porosity as control membrane. When NTMFs are used as the rate controlling membranes for the TTS patches, they show good zero order kinetics for a long period and within the therapeutic window. The quantity of drug or dose can also be regulated precisely without changing the surface area of the patch as required by conventional patches [22].

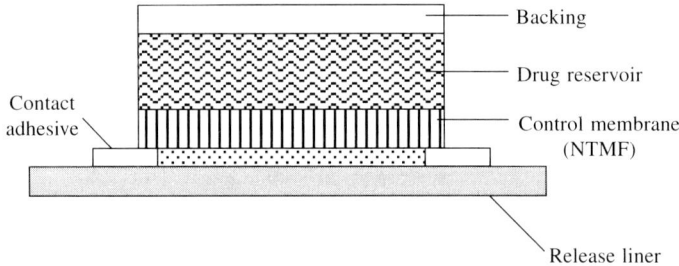

Figure 2.   A diagram showing different components of transdermal therapeutic system (TTS) including NTMF based control membrane

(c) Applications as diagnostic tools
An important aspect in medical diagnostics is the counting, sizing and separation of individual cells. It is in this field single pore membranes (SPMs) find their utmost utility. It has been found that certain cancers like those in the breast, colon and rectum release some malignant cell into the blood stream before the malignancy has metastasized. Thus by devising an appropriate SPM with a pore

diameter of 3 μm and filtering the blood samples through it may help in an early diagnosis of such cancers.

The characteristic feature of red blood cells is their elasticity, which sets them apart from malignant cells. Normal RBC is an oblate disk (doughnut shaped) with a diameter of 7.5 μm and a thickness in the range of 1–2 μm. A healthy RBC is extremely flexible and it can easily squeeze through the finer capillaries (3–5 μm) in the human body. However, many diseases of heart and circulatory system have been traced to alter deformability of the RBCs. Extensive research work is now in progress in many laboratories to study the flow behavior of blood in the capillary network.

Using single pore membranes, a special device has been developed to study the plasticity of RBCs [14]. Figure 3 shows a schematic view of such a device. The measurement of passage time of each RBC through cylindrical pore of this device can be easily performed. It has been observed that the infected cells loose their flexibility and become more plastic than the normal cells, hence they take longer passage time. The single pore membranes (SPMs) possess characteristic features to study several internal cell parameters in medical diagnostics [10, 13].

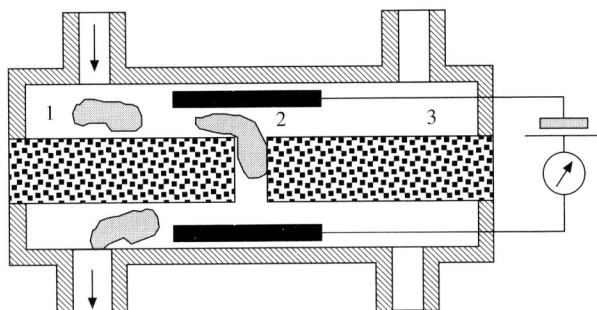

Figure 3. **A typical design of cell counter for characterization of individual red blood cells. (1) red blood cells, (2) electrodes and (3) the single pore membrane with pore diameter of about 4 μm.**

(d) Applications in counting and sizing of bacteria and viruses

Single pore membranes also find use in sizing and counting of bacteria. A typical example is the DeBlois Bean counter (3). This uses a single etched track to count and size small particles in an electrolyte. The two sides of an electrochemical cell are separated by a membrane containing just one micro-hole of well-known size. The resistance between the electrodes depends primarily on the conducting path through the hole. As a charged insulating particle enters, the resistance increases by an amount proportional to the volume of the particle. The velocity of the particle through the hole is a measure of its charge and so the two vital parameters namely size and charge of the particles can be used to characterize, and identify individual particles. This counter is specially useful for determining the size of viruses, which range from 15–450 nm in dimension and none can be examined under an optical microscope. With the help of DeBlois Bean counter one can easily count and characterize different bacteria and viruses.

(e) The development of bio-compatible polymer surfaces for cell growth

In recent years, the possibility about artificially produced hydrophilicity of polymer surfaces has attracted considerable attention in bio-medical applications. For instance, many implant devices developed as temporary or permanent replacements of biological organs use synthetic materials, including plastics and rubbers, which have to be exposed to tissues and body fluids. Ordinarily most polymers are of a hydrophobic nature. The surfaces of such synthetic materials need to be modified to make them hydrophilic and thus bio-compatible. Heavy ion irradiation of polymeric surfaces create nanometric damage trails characterized by active (ionized) sites. These trails can be filled by desired gases (oxygen) under high pressure. Also by suitable chemical etching of the irradiated polymers, one can enlarge the contact surface and enhance the roughness of the surface to facilitate the growth of cell cultures. Such products will be highly useful in cosmetic and pharmaceutical industries to test their products for toxicity on specialized cell-cultures grown on bio-compatible polymer surfaces instead of using test animals. Such activity will be highly welcomed by the voluntary environmental groups working against cruelty to animals.

(f) Detection of micro-organisms

The traditional methods used for detection of micro-organisms are based on slow growth leading to turbidment of nutrient medium. This process some times requires days. On the other hand, modern biological detection equipment offers faster measurement but involves high acquisition costs, therefore, becoming inaccessible to most clinical analysis laboratories. The examination and detection of micro-organisms by neutron radiography technique [21] using SSNTD (CR-39) requires no pre-cultivation. This method is simple, more accesible, efficient and low cost.

The micro-organisms are separated from the culture media and resuspended in a boron based lid solution. Later these are deposited in line detectors and at last submitted to a thermal neutrons beam (approx. 2.2×10^5 neutrons per square centimeter per second). The latent tracks registered by alpha particles coming from the nuclear reaction $^{10}B(n, \alpha)^7$ Li reactions are chemically etched and analysed by an optical microscope. The shape and size of individual micro-organisms and their agglomerations can be obtained as clear images [20].

(g) Applications in cellular radiation biology

Track detectors can provide valuable information on cell size from attenuation of the beam in living systems. First the cells grown on polymeric detector surface are exposed to heavy ions and then etched under suitable chemical conditions. From the etched surface one may get quantitative information about cell shape (mean thickness and monolayer fluctuations) and even a 'nuclear' cell image. Such information is especially valuable for a radiobiologist because it is affected by the same parameters as radiation dose. The use of track detectors in combination with targeted microbeam irradiation is expected to provide conclusive results on the biological effectiveness of single charged particle [7, 6].

Acknowledgments

The authors thank scientific and technical staff of GSI, Darmstadt (Germany) and NSC, New Delhi (India) for heavy ion irradiations.

References

1. Batzdorf, U., R.S. Knox, S.M. Pokress and J.C. Kennedy (1969) Membrane partitioning of the Rose-type chamber for the study of metabolic interaction between different cultures. *Stain Technology* 44: 71–74.
2. Cui, H.H., Shi-Cheng Wang, Ri-Sheng Wu and J. Zhou (1988) The production of nuclear track microfilters and their applications. *Nucl. Tracks Radiat. Meas.* 15: 767–770.
3. DeBlois, R.W. and C.P. Bean (1970) Counting and sizing of sub-micron particles by resistive pulse technique. *Rev. Sci. Instrum.* 41: 909–916.
4. Dey, M., J. Raju, S. Ghosh and K.K. Dwivedi (1993) Development and characterization of polycarbonate microfilters. *Nucl. Tracks Radiat. Meas.* 22: 907–908.
5. Durante, M., G.F. Grossi, M. Pugliese, L. Manti, M. Nappo and G. Gialanella (1994) Single charged-particle damage to living cells: a new method based on track-etch detectorsw. *Nucl.Instrum.Meth.* B94: 252–258.
6. Durnate, M., G.F. Grossi, M. Pugliese and G. Gialanella (1996) Nuclear track detectors in cellular radiation biology. *Radiat. Meas.* 26: 179–186.
7. Dwivedi, K.K. (1996) Special devices for bio-medical and radiobiological applications. *ISRB News Lett* 3: 3–4.
8. Dwivedi, K.K. (1997) New dimensions of heavy ion nuclear research. *J. Indian Chem. Soc.* 74: 357–366.
9. Fink, D., HMI. Berlin, Germany (Private Communications)
10. Fisher, B.E. and R. Spohr (1983) Production and use of nuclear tracks: imprinting structure on solids. *Rev.Modern Phys.* 55: 907–948.
11. Heise, S., K.K. Dwivedi, P. Vater, R. Brandt and C. Dankmeyer (1993) On the development of polypropylene (PP) micro-filters. *Nucl. Tracks Radiat. Meas.* 22: 909–910.
12. Horwitz, D.A. and M.A. Garrett (1971) Use of leukocyte chemotaxis *in vitro* to assay mediators generated by immune reactions. *J. Immunology* 103: 649–655.
13. Roggenkamp, H.G., H. Kiesewetter, R. Spohr, U. Dauer, L.C. Busch (1981) Production of Single-Pore membranes for the measurement of Red Blood Cell Deformability. *Biomedizinische Technik* 26: 167–169.
14. Roggenkamp, H.G., H. Kiesewetter, T. Schmeink and H.G. Hollweg (1982) Ein Neues Verfahren zur Quantifizierung der Verformbarkeit von Erythrocyten, Biomedizinische Technik., 27: 162.
15. Seal, S. (1964) A sieve for the isolation of cancer cells and other large cells from the blood. *Cancer.* 17: 637–643.
16. Song, J., P. From, W. Morrissey and J. Sams (1971) Circulating cancer cells: pre- and post-chemotherapy observations. *Cancer* 28: 553–561.
17. Spohr, R. and J.Vetter, GSI, Darmstadt, Germany (Private Communications).
18. Spohr, R. (1990) *Ion Tracks and Microtechnology – Principles and Applications,* Vieweg Verlag, Braunschweig.
19. Vater, P. (1988) Production and applications of nuclear track microfilters. *Nucl. Tracks Radiat. Meas.* 15: 743–749.

20. Wacha, R., V.R. Crispim, C. Lage and J. D'Arc R. Lopes (2000) Neutron Radiography applied to microorganisms detection. *Radiat. Meas.* 32: 159–162.
21. Walker, J. (1975) Radiography with Neutrons, In: Proceedings of the Conference Radiography with Neutrons, pp. 1–6, University of Birmingham, London.
22. Wu, R., J. Zhou and W. Ke (1993) Application of new nuclear track microporous membrane in transdermal therapeutic system (TTS), *Nucl. Tracks Radiat. Meas.* 22: 937–939.

Radiobiology and Bio-Medical Research
Edited by K.P. Mishra
Copyright © 2004 Narosa Publishing House, New Delhi, India

27. Recent Advances in Neurosciences: Their Therapeutic Relevance

P.N. Tandon

National Brain Research Centre, New Delhi 100 067

Neuroscience research is a continuum of study of the nervous system from the molecular to the behavioural level. It encompasses the body of research directed towards understanding the molecular, cellular, intercellular processes mediated through electrochemical signals, in the nervous system, integrated to subserve behavior. Rapid developments in diverse fields like molecular biology, immunology, genetics and biotechnology have helped in understanding the signaling process involved in receiving, analyzing, storing and processing information gathered both from the external and internal environment.

Ultimate Goal of Neurosciences

- To know the human brain
- To protect the human brain
- Treat the diseased brain
- To recreate the human brain

"Ultimately to unravel the mystery of mind."

Therapy

Information gained from basic research have already provided promising leads for development of better therapy. Intensive research on the mechanism of interneuronal communication has provided valuable information on a variety of neurotransmitters, neuromodulators, neurohormones and neurotrophic factors involved in growth, development and health of the neurons. The study of their synthesis, storage, release, degradation and recycling not only provides a fascinating picture of the chemistry of the brain but also new insights regarding their role in pathogenesis of many brain and mental disorders. This knowledge has provided valuable information on possible therapeutic targets and modeling of new drugs.

Efforts to develop new drugs to treat mental illnesses, behavior disorders and neurodegenerative disorders, rely in large part on detailed knowledge of receptors, ion-channels, transporters, membrane and protein structure of the involved molecules. Some of these are already in various stages of experimental and clinical trials. For example, considering that GABA is deficient in some forms of epilepsy, drugs targeted on enzymes involved in synthesis of GABA, proteins that bind GABA or enzymes that break down GABA have been developed.

Similarly, the discovery of a new molecule called Nurr 1, which is believed to play a critical role during embryonic development of dopaminergic neurons raises the possibility that boosting or restoring Nurrl activity in failing nerve cells may delay or prevent the onset of Parkinsonian symptoms.

Neurotransmitters—Receptors in Brain Disorders

A major finding over the past decade has been that the human brain uses only a few small molecules as neurotransmitters, but uses a huge diversity of different neurotransmitter receptors which send very different signals to the receiving nerve cell. The diversity of chemical coding accessible to nerve cells is very large and complex. Same neuron may synthesize and liberate several different chemical messengers subserving specific functions. Chemical maps of the brain are now becoming available.

Hundreds of neurotransmitter receptor genes have been isolated, cloned and studied in detail. This has paved a way for the pharmaceutical industry to develop new very specific drugs to target specific receptors and nerve networks raising the hope for treatment of mental illness and neurodegenerative diseases without the unpleasant side-effects associated with the currently used drugs.

Neurotransmitters involved in Brain Disorders

Disorder	Neurotransmitter/receptor
• *Head injury*	Excitatory Neurotransmitters: Acetyl Choline, Glutamate (Muscarinic/NMDA receptors)
• *Cerebral Ischaemia*	Glutamate (NMDA, AMPA) Kainate-receptors
• *Parkinson's Disease*	Dopamine, D1>D2 receptors
• *Alzheimer's Disease*	Cholinergic
• *Schizophrenia*	Dopaminergic D4 $\propto$ 2 5HT 2A/2C
• *Depression*	Serotonin
• *Panic Disorder*	Noradrenaline, $\propto$ 2

Ongoing or Initiated Clinical Trials:

Neurotrophic factor	Disease
• Nerve growth factor	Diabetic neuropathy Taxol Neuropathy Compressive neuropathy AIDS-related neuropathy Alzheimer's disease
• Brain-derived growth factor	Amyotrophic lateral sclerosis
• Neurotrophin 3	Large fiber neuropathy
• Insulin-like growth factor	Amyotrophic lateral sclerosis Vincristine neuropathy Taxol neuropathy
• Ciliary neurotrophic factor	Amyotrophic lateral sclerosis

Neuro protective: Under trial

- *Aptiganel*   NMDA receptor, ion channel blocker
- *ZD 9379*   NMDA receptor, glyctue site blocker
- *GV 150–526 A*   NMDA receptor,
- *SNX III*   Ca Channel blocker
- *Lobeluzole*   Interferes with nitric oxide's effects
- *Tirilazad*   Free Radical scavenger

Possible Therapeutic Applications of Neuropeptides

Peptide	Therapeutic application
• Vasopressing, [Arg^8 – vasopressin, AVP]	*Memory*
• Adrenocorticotropin ACTH (human)	*Memory, attention, Alzheimer's disease (AD)*
• Melanocyte-stimulating hormone, MSH	*Attention, fever*
• Melanostatin MIF-1	*Depression*
• Thyrotropin-releasing hormone, TRH	*Depression, AD*
• β-endorphin (human)	*Analgesia*
• Enkephalin (Met^5 –or Leu^5 –enkephalin)	*Analgesia*
• Kyotorphin	*Analgesia*
• Luteinising hormone-releasing hormone (LHRH)	*Sex hormone regulation*
• Cholecystokinin, CCK [8]	*Eating Disorder, schizophrenia*
• Delta sleep-inducing peptide, DSIP	*Insomnia*

Neurotrophins and Brain Insults

Epileptic, hypoglycaemic, ischaemic and traumatic insults to the brain induce marked changes of gene expression for the neurotrophins, nerve growth factor, brain-derived neurotrophic factor and neurotrophin-3 and their high-affinity receptors, TrkB and TrkC, in cortical and hippocampal neurones. Release of glutamate and influx of Ca^{2+} are the most important triggering factors. The major hypotheses for the functional effects of the insult-induced neurotrophin changes are protection against neuronal damage and stimulation of sprouting and synaptic reorganization. In-depth insight into the regulation and role of the neurotrophins after brain insults should increase our understanding of pathophysiological mechanisms in, for example, epileptogenesis and cell death and could lead to new therapeutic strategies.

Possible delivery systems for neurotrophic factors

- Systemically (only some disorders)
    Intra-cerebroventricular (ICV) pump system
    Intraparenchymal injection
    Installation of polymers with slow release

- 'Ex-vivo' gene therapy:
    grafting of genetically altered cells
    encapsulated cells (immuno-isolation)

Future application

— 'in-vivo' gene therapy (adenovirus–or Herpes virus-based vectors)
— molecules crossing blood-brain barrier
— active peptides, pantrophic molecules
— drugs with indirect effects on receptors

Molecular Genetics

The decade has seen an explosion in our knowledge of the molecules of the brain and how they function. The revolution in genetics and recombinant DNA technology has led us to genes that control the development and functions of the nervous system as well as genes whose mutations causes brain disease. Progress in structural biology has given us three-dimensional structures of many of the proteins encoded by these genes, offering new possibilities for drug development. To give one example let us consider Alzheimer's disease.

Molecular Genetics of Alzheimer's Disease

Molecular genetic studies of autosomal dominant familial Alzheimer's disease (FAD) has led to the identification of four distinct genes associated with this disorder: APP gene on chromosome 21, Preseniline 1 gene on chromosome 14, Preseniline 2 gene on chromosome 1 and Apolipoprotein gene on chromosome 19. Another gene on chromosome 12 (1 gene for macroglobulin and the low density lipoprotein related protein LRP1) has also been proposed as reasonable candidate. Further studies based on this information suggest that abnormalities in the processing of APP protein is a common feature in most forms of AD, resulting in over-production of A-peptide. Currently attempts are being made to design treatments, which will modulate the production and/or disposition of A-peptides.

Some possible Drugs for preventing or treating Alzheimer's

Drugs	Activity	Mechanism
Cognex Aricept nurons	Acetylcholinestrase inhibitor	Compensate loss of cholinergic
Ampakines	Activity of AMPA Rec.	Improved memory by enhancing LTP
Vitamin E	antioxidant	Protects against free radicals
Premarin	Female Hormone	Promotes neural survival
Nerve Growth Factor	Maintains Ch 1. Neurons	Promotes neural survival
Calcium Channel	Inhibits Ca^{++} entering Neurons	Calcium Toxicity
Cholestrol Lowering D	Lower apo E4 cone	Prevents apoE4 Toxicity
Protease	Prevents β-amyloid	Prevents neuronal loss due
Inhibitors	production	to β-amyloid toxicity
Prednisone	Anti inflammatory	Prevents inflammatory damage to neurons

Jean Marx: The Best of Science: Neuroscience (1999).

Researchers have now isolated the enzymes or γ or β secretase that make-amyloid that is believed to be responsible for neuronal degeneration associated with (or responsible for) Alzheimer's disease. It is hoped that if amyloid is the destructive agent in Alzheimer's disease, drugs that target its production could slow or even reverse the disease.

β amyloid is formed by cleavage of APP (Amyloid Precusor Protein) by enzyme β-secretase and/or γ-secretase which have now been isolated. Instead of attempting to isolate the enzyme (which is difficult) scientists of Amegen Company detected a single gene that raises β-amyloid production. The protein encoded by this gene BACE (Betasite APP-clevage enzyme) proved to have all the properties of b-secretase. In habiting this enzyme decreased β-amyloid production by cultured cells.

- Several companies have now developed compounds that inhibit APP cleavage thus preventing formation of β-amyloid containing plaques.
- Evidence is now produced to suggest that presenilin 1, implicated in some formes of inherited Alzheimer's may intact by γ-secretase.
- Having both β-γ-secretase a number of companies now note that BACE, is similar to the HIV protease in the AIDS virus and many compounds have already been developed to inhibit that enzyme.

One of the mechanisms responsible for neural damage following cerebral trauma, iscaemia and spinal injury that has been extensively investigated during the **decade of the brain** has been the glutamate induced excitotoxicity—resulting in large amounts of calcium to enter the nerve cells. This calcium overload results in generation of free radicals capable of destroying many cellular structures leading to cell death. In addition more recently it has been demonstrated that besides destroying neurons, it can also damage oligodendrocytes and consequently the myelin. The damage mediated by glutamate and calcium overload may be augmented by zinc. Therapeutic strategies based on this knowledge are undergoing experimental and clinical trials.

Radiobiology and Bio-Medical Research
Edited by K.P. Mishra
Copyright © 2004 Narosa Publishing House, New Delhi, India

28. A Novel SCID Biotechnology for Biomedical Studies in Human

T. Nomura, H. Nakajima, T. Hongyo, H. Ryo, L.Y. Li,
M. Kurooka, R. Baskar, M. Syaifudin, X.E. Si, M. Maeda,
R. Kaba, K. Mori, Y. Fukudome, J. Y. Koo, K.M.Y.Wani,
Y. Kitagawa, R. Tsuboi, K. Hiramatsu and F. Matsuzuka

Graduate School of Medicine, Osaka University, 2–2 Yamada-Oka, Suita,
Osaka, Japan

Introduction

In the development of human culture, especially after the Industrial Revolution, enormous numbers of human-made physical and chemical agents have been produced. These toxic agents come to humans directly or after the biological concentration, whereas humans have never been exposed to those agents. Humans have been exposed to solar ultraviolet light B (UVB) by the depletion of stratospheric ozone. These environmental hazards to humans have been evaluated by the epidemiological survey, animal experiments and/or in vitro studies. For the exact estimation of the risk to human health, however, establishment of the direct methods with human organs and tissues is extremely important. For this purpose, a novel biotechnology has been established with severe combined immunodeficient (SCID) mice which cannot reject human organs and tissues immunologically [1–7].

This biotechnology also enables us to study: (1) morphology and function of human organs and tissues by the reconstruction of the human organ system in SCID mice, (2) experimental therapy of human diseases, including medicinal, immuno- and gene-therapy, (3) mutagenesis and carcinogenesis in human, including toxicity of new drugs, and (4) development of human embryos.

Improvement of SCID Mice

Scid mutation was found by Bosma in 1983 [1]. SCID mice were defective of VDJ recombination in the development of lymphocytes, showing deficient T cell and B cell function, while athymic nude mice lack T cell function, but possess normal B cell function. However, normal T and B cells appear spontaneously in 13.6–23.0 percent of the original C.B17-*scid* mice ('leaky' SCID mice) and 31.7–58.8 percent of them die of leukemia early in life (2). These characteristics give serious disadvantage to maintain human organs and tissues for long period. In fact, a majority of mouse skin allografts (> 70 percent) were rejected in leaky SCID mice.

In 1986, we started selective brother-sister inbreeding of homozygous C.B17-*scid/scid* mice which show undetectable IgG and IgM (< 1 μg/ml), and also introduced other immunodeficient genes into SCID mice. As shown in Fig. 1, 8.4 to 20.0 percent of C.B17-*scid* mice showed unusually high serum level of IgG ($\geq$ 25 μg/ml) until F_5 generation. The incidence of such leaky SCID mice decreased after F_6 by selective inbreeding of IgG and IgM undetectable C.B17-*scid* mice. Incidence of lymphocytic leukemia was also extremely high (7.7–16.5 percent) until F_7 generation, but decreased significantly after then by selective brother-sister inbreeding of IgG and IgM undetectable C.B17-*scid* mice (Fig. 2) (40/302, 13.2 percent at F_{3-7} *vs* 46/1569, 2.9 percent at F_{8-23}, p < 0.01). We designated this new inbred strain as C.B17/N-*scid* after F_{20} generation. Among 443 C.B17/N-*scid* mice at F_{39-41} (early 1999), none of them (zero percent) are leaky and 13 C.B17/N-*scid* mice (2.9 percent) died of leukemia.

Leaky SCID was observed very often, when *scid* gene was introduced into a different background. Since SCID mice possess normal natural killer (NK) cell and macrophage function, *scid* gene was introduced into other inbred strains of mice, such as C57BL/6J, C57BL/6J-bg^J , C3H/HeJ, and B6C3Fe-*a/a*-Csfm^{op} mice by cross-back cross mating. It resulted in extremely high incidence of leaking of IgG (but not IgM) in *scid* homozygous mice. However, leaky SCID disappeared by the brother-sister inbreeding of *scid*-homozygous male and female (Fig. 1).

In the improved SCID mice, not only benign and malignant tumors, we succeeded to maintain normal human organs and tissues for a long period, reserving morphology and function [11, 3], and human embryos also develop well and grow in the improved SCID mice.

Transplantation Procedure

Normal human organs and tissues were treated with high concentration of antibiotics (50,000 units/ml penicillin G, 25 mg/ml panipenem, 25 mg/ml betamipron, and 50 mg/ml streptomycin sulfate) for 15 minutes and washed repeatedly with the medium containing antibiotics. Then, human tissues were transplanted subcutaneously into the SCID mice. When mice become moribund, human tissues were removed and transplanted to another set of SCID mice repeatedly to maintain human tissues beyond several mouse generations. Though SCID mice will die, human organs and tissues can survive for a long period. At the time of transfer, biopsy of human tissues were carried out to detect morphological and molecular changes. It is possible to treat human tissues with radiation and/or chemicals during these processes.

All experiments were carried out in the germ-free or SPF room of the barrier section in the Institute of Experimental Animal Sciences, Osaka University following the Osaka University Guideline for Animal Experimentation. Human tissues resected from the patients for surgical treatment of cancer were used with their permission for heterotransplantation in the SCID mice. Only the human tissues free of mycoplasma, human hepatic B and C antigens/antibodies, adult T-cell leukemia, HIV, and Wasserman reaction were accepted into the SPF room of the barrier [1–7].

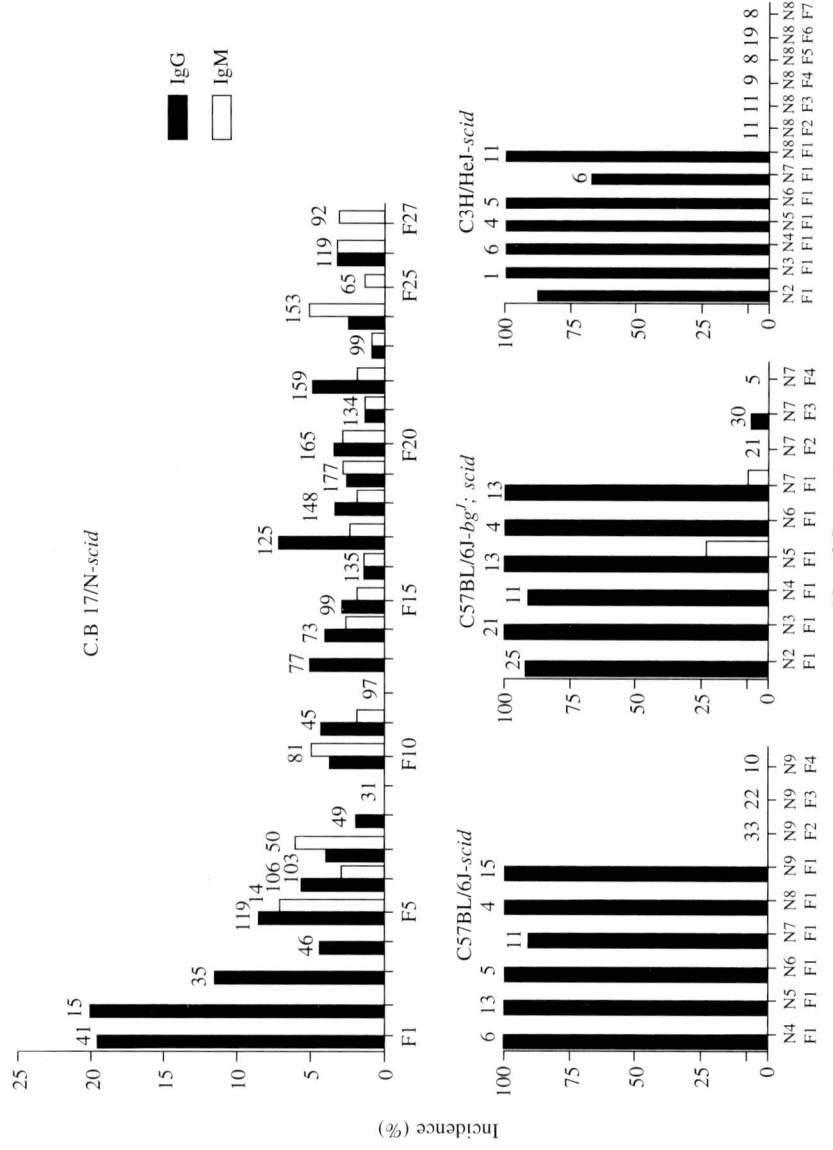

Figure 1. Incidence of leaky SCID mice in the process of inbreeding of C.B17-*scid* mice and genetic introduction of *scid* gene into C57BL/6J, C57BL/6J-*bg^J* and C3H/HeJ mice. SCID mice showing serum IgG and IgM higher than 25 μg/ml were designated as leaky SCID mice. Details for measurements of IgG and IgM by ELISA were given in ref. 2 and 4. Figures on the histogram indicate numbers of SCID mice.

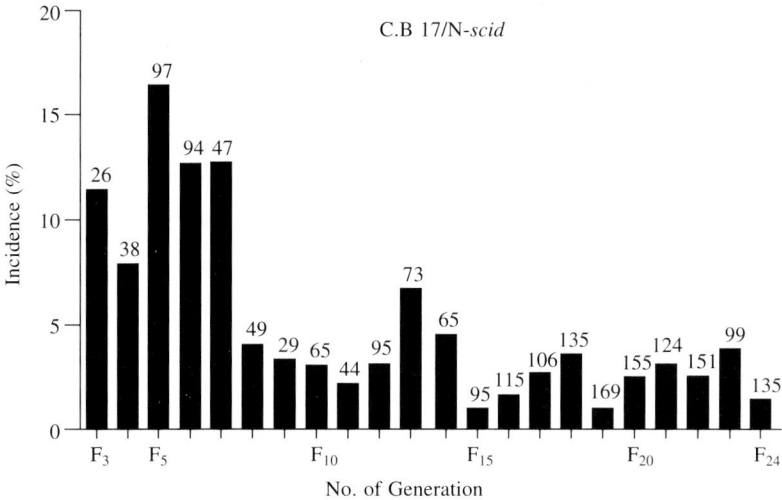

Figure 2. **Incidence of thymic lymphocytic leukemia in the process of selective inbreeding of C.B17-*scid* mice. Figures on the histogram indicate numbers of SCID mice.**

Human Benign and Malignant Tumors in SCID Mice

All human malignant tumors so far tested can grow more rapidly in the improved SCID mice than athymic nude mice [1–4, 10], while about half of human malignant tumors could not grow in nude mice. Furthermore, these transplanted human malignant tumors metastasized to distant organs, while metastases have never been observed in nude mice [8, 9]. Human benign tumors also could grow in the improved SCID mice for long periods (~ 3 years) over several mouse generations (Table 1). Microscopic structures of transplanted human benign and malignant tumors were well maintained in the SCID mice [1–4], and transplanted human tissues were also stable in SCID mice, for example, human chorionic gonadotropin and parathyroid hormone were secreted in the urine and blood of C.B17-*scid* mice after transplantation of hydatidiform mole (unpublished data) and parathyroid adenoma [7]. Consequently, experiments were extended to the transplantation of normal human organs and tissues to the improved SCID mice, C.B17/N-*scid*.

Maintenance of Normal Human Organs and Tissues

Normal human organs and tissues such as skin, thyroid gland, lung, liver, gastrointestinal tract, endometrium and bone marrow cells and also precancerous lesions are well maintained in C.B17/N-*scid* mice for long periods over several mouse generations (Table 1). Histological patterns of these transplanted human organs and tissues are also well maintained in the C.B17/N-*scid* mice [10]. However, human bone marrow cells could not survive longer than 6 months in our previous study with C.B17-*scid* mice from F_2 to F_7 generation [10], when survival was confirmed by the serum level of human IgG. However, higher level of human IgG was detected when bone marrow cells were injected into the improved C.B17/N-*scid*; W^v doubly mutant mice than when injected into usual

Table 1. Survival and maintenance periods of normal and precancerous human tissues in SCID mice

	No.	Survival (%)	Max. period[a] (months)	No. of transfer
1. Normal human tissues				
Skin	42	100	28	7
Thyroid gland	49	100	34	10
Lung	13	100	22	8
Liver	7	100	14	4
Gastric mucosa	18	100	21	6
Colorectal mucosa	10	100	13	5
Endometrium	7	71.4	22	6
Bone marrow cell	19	84.2	< 6[b]	1
2. Benign human tumors				
Thyroid follicular adenoma	10	100	36	13
Parathyroid adenoma	15	83.3	13	3
Parotid pleomorphic adenoma	15	100	36	11
Papilloma (tongue, larynx)	5	100	34	7
Leukoplakia (tongue)	5	100	8	2
Hydatidiform mole	18	88.9	12	4
Endometrial Hyperplasia	18	83.3	14	5
Colorectal polyp	46	63.0	37	11
Pancreas adenoma	10	50.0	13	3

(a) Some continue.
(b) Identified by human IgG in the serum of SCID mice. Usual C.B17-*scid* mice were used.

C.B17-scid mice [10]. Major differences between transplanted bone marrow cells and solid human organs are the existence of supporting tissues in solid organs. The second transfer is very difficult in human bone marrow cells.

When radio-iodine, Na^{125}I, was injected intravenously into SCID mice with normal human thyroid xenograft, 20 to 27 percent of radioactivities were incorporated into transplanted human thyroid gland, and 6 to 11 percent in the mouse thyroid gland for 24 hours after ^{125}I injection. Thus, transplanted human normal thyroid gland is well incorporating iodine [3]. Human thyroid hormone (T3) from human thyroid tissues were detected in the blood of SCID mice with human thyroid gland, while it was not detectable in SCID mice without human thyroid gland transplantation. Level of T3 increased about three fold by the thyroid stimulating hormone (TSH) treatment to SCID mice. These results indicate that transplanted human thyroid gland shows normal function, that is, incorporation of iodine, continuous secretion of T3 and release of T3 by the TSH treatment.

Not only the adult human tissue, various organs of human embryos 10–11 weeks old were also transplanted into SCID mice with permission. Embryonic organs differentiate and grow rapidly in the improved SCID mice. For example, tiny finger (about 1 mm diameter) forms 5mm width finger with nail four mouths after transplantation. Internal organs also develop and grow rapidly, keeping and forming their organ specific morphology.

Mutagenesis and Carcinogenesis Studies in Human Tissues

Table 2 summarizes the effects of some organ-specific mutagens/carcinogens on the normal human tissue; N-amyl-N-methyl-nitrosamine (AMN) on esophagus, 1-methyl-3-nitroso-1-nitrosoguanidine (MNNG) on gastric and colon mucosa, urethane on the lung, X- and γ-rays on thyroid gland, or UVB on the skin. When human organs and tissues were treated with specific carcinogens/mutagens repeatedly for one to two years, *p53*, K-*ras*, *c-kit* mutations were induced in human tissues, and later cancers developed in the human skin and esophagus. Mutations were detected by the Cold-SSCP and direct sequencing [4].

Table 2. **Mutation and cancer induced in normal human tissues by organ-specific mutagenic and carcinogenic agents**

Organs	Mutagen and carcinogen	Mutation	Cancer
Esophagus	AMN	*p53*	+
Stomach	MNNG	K-*ras*, *p53*	−
Colon	MNNG	K-*ras*, *p53*	−
Lung	Urethane	K-*ras*, *p53*	−
Lung	Dioxin (TCDD)	K-*ras*	−
Thyroid gland	X-rays, TCDD	*p53*, *c-kit*	−
Skin	UVB + A	*p53*, *c-kit*, K-*ras*	+

Ultraviolet light exposure

Solar light has continuously influenced the development, metabolism, and so on of all organisms on the earth. For human beings, it is beneficial in general, but a part of solar light, UV radiation, can be harmful. Recently, the depletion of stratospheric ozone and the consequent increase in levels of genotoxic UVB (290–320 nm) in sunlight have led to fear of a rise in the frequency of skin cancer in human populations. Consequently, we maintained human skins in the improved SCID mice, and then irradiated daily with UVB for a long period (~2 years) to confirm the direct link between UVB and the development of human skin cancer, solar (actinic) keratosis, and molecular changes in *p53, ras* and *c-kit* genes. We used normal human skin at breast site of breast cancer patient and human foreskin of the penis of the phimosis patients with their permission, because these skins are rarely exposed to sunlight. Ultraviolet light shorter than 290 nm was cut-off by Kodacel cellulose film, and UVB and a spectrum which is similar to solar UV on the surface of the earth was given daily to human skin on the SCID mice; 0.8 J/cm^2 on Monday, Wednesday and Saturday, and 0.2 J/cm^2 on Tuesday, Thursday, Friday and Sunday. Minimum erythema dose for Japanese is 0.03–0.15 J/cm^2. After the daily exposure to UVB, human skin became irregularly thick and brownish, whereas no changes were observed in the mouse skin, simply because the mouse skin was covered by the hair. Microscopic feature showed typical actinic keratosis [11].

After the long continued UVB irradiation with about 100 J/cm^2 in total doses, ulcerated skin tumor developed sequentially in the brownish human skin and grew rapidly. Three squamous cell carcinomas and one undifferentiated skin cancer developed. UVB induced skin tumors were transplantable to C.B17-*scid/scid* mice, but not to their wild type C.B17–+/+ mice. Furthermore, DNA extracted from UVB induced skin tumor was amplified by the primers of human *p53* and K-*ras* genes, while DNA of the mice were not. Thus, it is confirmed immunologically and genetically that induced skin tumor is malignant and has originated in human skin. This is the first experimental success to induce malignant tumor in human tissues [6].

UVB-irradiated and unirradiated human skin pieces were examined for *p53*, K-*ras* and *c-kit* mutations. *p53* mutations were often detected in the UVB-irradiated human skin, especially in about 80 percent of UVB induced actinic keratosis and all of skin cancer, although K-*ras* and *c-kit* mutations were not so often in UVB-induced actinic keratosis skin. About 85 percent of *p53* mutations occurred at dipyrimidine sites. T-C transition at codon 242 (Cys-Arg) was specifically observed in both skin cancer and actinic keratosis, and double or triple *p53* mutations were observed in UVB-induced skin cancers. In two of three UVB induced human skin cancer, K-*ras* and *c-kit* mutations were also induced in addition to duplicated and triplicated *p53* mutations. Consequently, three to five mutations were accumulated in UVB induced skin cancer [11].

Radiation and radionuclide exposure

Thyroid gland is known to be very sensitive to radiation, and radio-iodine. Radio-therapy of thymoma patients caused thyroid cancer, and radioiodine exhausted by the Chernobyl catastrophe caused high incidence of thyroid cancer in exposed children. Thyroid gland is also the target of endocrine disrupters such as dioxins.

Normal human thyroid glands from the head and neck cancer patients were transplanted to the SCID mice. SCID mice with human thyroid glands were exposed weekly to X-rays (2 Gy) or ^{137}Cs γ-rays (1 Gy), or 0.4 MBq of Na^{131}I was injected subcutaneously in to SCID mice. With increasing doses of X- or γ-rays and repeated injection of ^{131}I, brownish thyroid glands became pale and very small, and necrotic parts increased histologically. Without radiation exposure, level of human thyroid hormone in the peripheral blood of SCID mice with human thyroid gland is same at 60 to 300 days after transplantation, indicating constant secretion of human thyroid hormone from transplanted human thyroid gland in unexposed controls. However, level of thyroid hormone decreased significantly after X- or ^{137}Cs γ-ray exposure. Results are compatible with macroscopic and histological findings.

High doses of X-rays (24 to 60 Gy) to human thyroid tissues produced *p53* mutation at codon 209 and *c-kit* mutation at codon 825, although lower doses did not. However, thyroid cancer has not yet developed, because thyroid gland of an old person is not sensitive to radiation.

Perspectives

Improvement in SCID mice enabled us to maintain normal and precancerous human organs and tissues for long periods. Although there remain some unsolved problems such as replacement of human bone marrow system, most human organs and tissues survive in mice with their normal morphology and function.

Not only to investigate the influence of radiation and environmental mutagens/carcinogens on the human health, establishment of SCID biotechnology with human tissues provides an invaluable experimental system for studying physiology, pathology, pharmacology and biochemistry of human organs and tissues, interaction between organ and organ, diagnosis and therapy/prevention of human diseases, and toxicity test of new drugs, without illegal and/or troublesome clinical trials.

References

1. Bosma G.C., R.P. Custer and M.J. Bosma (1983) A severe combined immunodeficiency mutation in the mouse. *Nature* 301: 527–530.
2. Bosma G.C., M. Fried, R.P. Custer and A. Carroll, D.M. Gibson and M.J. Bosma (1988) Evidence of functional lymphocytes in some (leaky) scid mice. *J. Exp. Med.* 167: 1016–1033.
3. Fukuda, K., H. Nakajima, E. Taniguchi, K. Sutoh, H. Wang, P.M. Hande, L.Y. Li, M. Kurooka, K. Mori, T. Hongyo, T. Kubo and T. Nomura (1998) Morphology and function of human benign tumors and normal thyroid tissues maintained in severe combined immunodeficient mice. *Cancer Letters* 132: 153–158.
4. Hongyo, T., G.S. Buzard, R.J. Calvert and C.M. Weghorst (1993) 'Cold-SSCP': a simple, rapid and non-radioactive method for optomized single-strand conformation polymorphism analyses. *Nucleic Acids Res.* 21: 3637–3642.
5. Inohara, H., T. Matsunaga and T. Nomura (1992) Growth and metastasis of fresh human benign and malignant tumors in the head and neck regions transplanted into scid (severe combined immunodeficiency) mice. *Carcinogenesis* 13: 845–849.
6. Inohara, H., T. Nomura, T. Hongyo, H. Nakajima, T. Kawaguchi, K.Fukuda, K. Sutoh and T. Matsunaga (1993) Reduction of leaky lymphocyte clones producing immunoglobulins and thymic lymphocytic leukemia by selective inbreeding of SCID (severe combined immunodeficiency) mice. *Acta Otolaryngol (Stockh)* 501: 107–110.
7. Kawaguchi, T., H. Nakajima, T. Hongyo, K. Fukuda, E. Taniguchi, K. Sutoh, H. Wang, P. Hande, L.Y. Li, M. Kurooka, T. Iwasa, N. Kurokawa, R. Nezu, M. Miyata, H. Matsuda and T. Nomura (1997) Consecutive maintenance of human solitary and hereditary colorectal polyps in SCID mice. *Cancer Detection and Prevention*, 21: 148–157.
8. Nomura, T., Y. Takahama, T. Hongyo, H. Inohara, H. Takatera, H. Fukushima, Y. Ishii and T. Hamaoka (1990) SCID (severe combined immunodeficiency) mice as a new system to investigate metastasis of human tumors. *J. Radiat. Res.* 31: 288–292.
9. Nomura, T., Y. Takahama, T. Hongyo, H. Takatera, H. Inohara, H. Fukushima, S. Ono and T. Hamaoka (1991) Rapid growth and spontaneous metastasis of human germinal tumors ectopically transplanted into scid (severe combined immunodeficiency) and scid- nude[streaker] mice. *Jpn. J. Cancer. Res.* 82: 701–709.

10. Nomura, T., H. Nakajima, H. Inohara, K. Fukuda, T. Kawaguchi, H. Wang, K. Sutoh, H. Sugiyama and Y. Oka (1993) Transplantation of human organs into scid mice. *Hematol & Oncol* 26: 461–469 (in Japanese).

11. Nomura, T., H. Nakajima, T. Hongyo, E. Taniguchi, K. Fukuda, Y.L. Li, M. Kurooka, K. Sutoh, T.M. Hande, T. Kawaguchi, M. Ueda and H. Takatera (1997) Induction of cancer, actinic keratosis and specific *p53* mutations by ultraviolet light B in human skin maintained in SCID mice. *Cancer Res.* 57: 2081–2084.

12. Phillips, R.A., H.A.S. Jewett and B.L. Gallie (1989) Growth of human tumors in immune deficient scid mice and nude mice. *Curr. Topics Microbiol Immunol* 152: 259–263.